T0361897

ROUTLEDGE LIBRARY EDITIONS: LOGIC

Volume 7

QUANTIFICATION THEORY

QUANTIFICATION THEORY

J. A. FARIS

LONDON AND NEW YORK

First published in 1964 by Routledge & Kegan Paul Ltd

This edition first published in 2020
by Routledge
2 Park Square, Milton Park, Abingdon, Oxon OX14 4RN

and by Routledge
52 Vanderbilt Avenue, New York, NY 10017

Routledge is an imprint of the Taylor & Francis Group, an informa business

British Library Cataloguing in Publication Data
A catalogue record for this book is available from the British Library

ISBN: 978-0-367-41707-9 (Set)
ISBN: 978-0-367-81582-0 (Set) (ebk)
ISBN: 978-0-367-41807-6 (Volume 7) (hbk)
ISBN: 978-0-367-85397-6 (Volume 7) (ebk)

Publisher's Note
The publisher has gone to great lengths to ensure the quality of this reprint but points out that some imperfections in the original copies may be apparent.

Disclaimer
The publisher has made every effort to trace copyright holders and would welcome correspondence from those they have been unable to trace.

QUANTIFICATION THEORY

BY

J. A. Faris

ROUTLEDGE & KEGAN PAUL
LONDON

First published 1964
by Routledge & Kegan Paul Ltd
Broadway House, 68–74 *Carter Lane*
*London, E.C.*4

Printed in Great Britain
by Latimer Trend & Co Ltd., Plymouth

CONTENTS

ABBREVIATIONS

Add	Rule of addition
CI_F	Implication corresponding to F
Conj	Rule of conjunction
CP	Rule of conditional proof
EG	Rule of existential generalization
EI	Rule of existential instantiation
EXT	Rule of extensionality of identity
HS	Rule of hypothetical syllogism
ID	Rule of identity
Int	Rule of interchange
MP	*Modus ponens*
MT	*Modus tollens*
MTP	*Modus tollendo ponens*
Pr	Premiss rule
QFE^n	Quantifier-free equivalent with respect to universe of n individuals
SD*	Rule of singular dequantification
Simp	Rule of simplification
SQ*	Rule of singular quantification
SQFE	Simplified quantifier-free equivalent
TF	Rule of truth-functional deduction
TFL	*Truth-Functional Logic*
UG	Rule of universal generalization
UGC	Unanalysed general component
UI	Rule of universal instantiation

CHAPTER ONE

1. Introduction. Quantification theory is concerned fundamentally with general arguments. We begin by explaining what we mean by this term. In much of our reasoning we make use of the ideas expressed by the words or phrases: *all*, *every*, *each*, *any*, *some*, *at least one*, *not all*, *not one*, *none*. Let us refer to these expressions and to others which are similar in meaning and use as *generalizers*. A *distinctively general argument* may be defined as any argument the force of which depends on the meaning and use of one or more generalizers. For example the following simple argument the force of which depends on the generalizers *every* and *some* is a distinctively general argument.

> Every applicant will be employed; some refugees will be applicants; therefore some refugees will be employed.

Distinctively general arguments form the main part of the field of quantificational logic. However, it would not be satisfactory to define quantificational logic as being concerned solely with distinctively general arguments; so we must try to state the position more exactly. In doing so it will be necessary to refer to truth-functional logic. We assume in the present book, which is indeed written more or less as a companion volume to my earlier monograph *Truth-Functional Logic*,[1] a reasonable acquaintance with this subject.

It is desirable to regard quantificational logic as including truth-functional logic. It follows of course that it is applicable to any arguments to which truth-functional

[1] *Truth-Functional Logic* will be referred to hereafter as TFL.

logic is applicable. Let us refer to any argument to which truth-functional logic is applicable as a *truth-functional argument*[1] and let us say that an argument is a general argument if and only if it is either distinctively general or truth-functional. We may now say that the field of quantificational logic is the class of general arguments. Our purpose is to show how truth-functional logic may be extended to form a system, quantificational logic, which is applicable to any general argument.

There are of course other systems of logic which deal with distinctively general arguments. One such system is the traditional syllogistic logic; another is the Boolean algebra of classes. However, neither of these is fully comprehensive. To get a true comparison we need to consider what the scope of each of these systems would become if it were supplemented by truth-functional logic. We shall say that a system of logic is adequate to a certain field if it provides non-intuitive methods by which we may attempt to test the validity of any argument within the field. Now the traditional syllogistic, even when supplemented by truth-functional logic, is adequate only to a small section of the field of distinctively general arguments. Similarly a supplemented Boolean algebra of classes is adequate only to a part of the field. The interest of quantificational logic, as compared with these earlier and related disciplines, is that it is adequate to the whole field of general, including distinctively general, arguments.

2. Direct and indirect evaluation. We shall speak of *evaluating* an argument. By this we mean determining whether an argument is valid or invalid. Any particular system of formal logic provides methods of evaluating arguments of a certain kind. In general the methods of a system of formal logic apply directly only to arguments

[1] The scope of truth-functional logic is discussed in TFL, chapter V.

which have a certain standard form. We can make use of these methods to evaluate an argument Z only if Z is already in the standard form or if we are able to find some argument Z' which is in the standard form and is equivalent to Z in the sense that the premisses of Z' can be true if and only if the premisses of Z are true and the conclusion of Z' can be true if and only if the conclusion of Z is true. When two arguments Z and Z' are equivalent in this sense we may say that Z is expressible in standard form as Z'. If an argument Z is in the form which is standard for a particular system of logic the methods of that system may be applied to it directly. If Z is not in the standard form the methods of the system in question may yet be used indirectly, provided that Z can be expressed as Z' where Z' is in the standard form: the methods of the system are in this case applied directly to Z' and from the result the evaluation of Z itself immediately follows.

Quantificational logic is a particular system of formal logic the methods of which are applicable directly to arguments of a certain standard form which will here be called *quantificational form*. All general arguments can be expressed in quantificational form and may thus be evaluated by quantificational logic. The evaluation is indirect in the sense explained in the last paragraph. To evaluate a general argument G we first express it as an argument Q in quantificational form; we then use quantificational methods to evaluate Q and when this has been done we are able immediately to evaluate G. It can thus be seen that to be able to use quantificational logic to evaluate general arguments we need to be able to do two things; first, to express general arguments in quantificational form; second, to apply the methods of quantificational logic. What we are here calling quantification theory may accordingly be regarded as having two main elements:

(1) The study of quantificational form and of the way

in which general arguments may be expressed in quantificational form.

(2) Quantificational logic proper, an account of the methods by which arguments in quantificational form may be evaluated.

The study of quantificational form will be our main concern in the rest of this chapter and in chapter three, sections 4 and 5; and broadly speaking quantificational logic proper will be dealt with in the remaining parts of the book.

We proceed now to give a preliminary account of quantificational form and an explanation of how general arguments may be expressed in this form.

3. Quantifier-matrix form of singular statements. Any statement[1] which is premiss or conclusion of a general argument and is not truth-functionally compound is either a general statement or a non-general statement. A general statement is normally expressed in quantificational form by means of a formula consisting of two main elements, a quantifier and a matrix. A non-general statement is not normally expressed in a quantifier-matrix form. However, although this is so, it will be convenient to begin here by taking a simple non-general statement and showing in stages how it may be transformed into a statement with the distinctively quantifier-matrix structure. The quantificational expression of general statements can then be explained as a natural development from the quantificational expression of non-general statements.

[1] The word 'statement' is used throughout the present book in the sense in which 'proposition' is used in TFL. I hope that what this sense is will become sufficiently clear in the contexts. I may add that I do not now regard what is said about 'proposition' on page 2, lines 9 to 16, of TFL as giving a satisfactory account of the use of the word in that book.

form of singular statements

Consider the following statement:

(1) Nero succeeded Claudius.

In some contexts it would be natural to express this in this form:

(2) Of Nero it is true that he succeeded Claudius,

which we may regard as consisting of two parts: a prefix, *Of Nero it is true that* and a matrix *he succeeded Claudius*. Let us now substitute a different prefix, *Nero is such that*. This gives us:

(3) Nero is such that he succeeded Claudius.

(3) might normally be taken to differ slightly in meaning from (1) and (2); it might be supposed not merely to mean that Nero succeeded Claudius but also to carry the implication that he was the sort of person who might have been expected to succeed Claudius and perhaps even the implication that he still exists. We stipulate that in what follows the phrase *is such that* is to be understood as being entirely without any additional implications of this sort. The prefix *Nero is such that* is to be understood as having exactly the same sense as *Of Nero it is true that*. Thus (1), (2) and (3) all have the same meaning.

We now transform (3) in a different sort of way. Using a construction found in verse:

Lars Porsena of Clusium By the Nine Gods he swore—

and common enough in colloquial speech, though frowned upon in literary prose, let us instead of (3) write:

(4) Nero he is such that he succeeded Claudius,

which we may regard as consisting of two parts, the prefix *Nero he is such that* and the matrix *he succeeded Claudius*.

Just as (1) referred to both Nero and Claudius so does (4). Let us now renumber (4) as (1′) and apply to it in respect of *Claudius* the same transformations as we applied to (1) in respect of *Nero*. We obtain in succession:

1.3. *Quantifier-matrix*

(1′) Nero he is such that he succeeded Claudius.

(2′) Of Claudius it is true that Nero he is such that he succeeded him.

(3′) Claudius is such that Nero he is such that he succeeded him.

(4′) Claudius he is such that Nero he is such that he succeeded him.

We may regard (4′) alternatively as consisting of the prefix *Claudius he is such that* and the matrix *Nero he is such that he succeeded him* or as consisting of the two prefixes *Claudius he is such that* and *Nero he is such that* and of the matrix *he succeeded him.* Unfortunately (4′) taken by itself is ambiguous: it could be interpreted as meaning either that Nero succeeded Claudius or that Claudius succeeded Nero. (4′) contains four occurrences of the third person pronoun. At the first occurrence it obviously refers to Claudius and at the second to Nero but it is not clear what the references are at the third and fourth occurrences. We may remove the ambiguity by labelling the pronouns; let us write (x) after the first and fourth pronouns and (y) after the second and third. This gives us:

(5′) Claudius he (x) is such that Nero he (y) is such that he (y) succeeded him (x).

It is an obvious convention that pronouns marked with the same letter have the same references. Given this convention (5′) can mean only that Nero succeeded Claudius. However, there is really no need to have pronouns in addition to the letters; the letters may serve instead of pronouns. Thus instead of (5′) we may write:

(6′) Claudius x is such that Nero y is such that y succeeded x.

In (6′) the prefixes *Claudius x is such that*, *Nero y is* *such that* are examples of what we will call *singular quantifiers*.[1] We may write them for short as (*Claudius x*) and (*Nero y*) respectively. Using this notation we obtain instead of (6′):

(7′) (Claudius *x*) (Nero *y*) *y* succeeded *x*.

Before we go further it will be useful to draw attention to certain features of this transformation. In the first place (1) *Nero succeeded Claudius* is an example of a *singular statement*, i.e. of a statement containing at least one proper name. Proper names normally occur in the body of a statement. Singular quantification, i.e. the expression of a singular statement in a form involving singular quantifiers, is a device whereby we may exhibit separately, on the one hand the proper names which a statement contains and on the other hand what is said about the individuals which the proper names denote. Thus in our example (7′) the proper names, *Claudius* and *Nero*, are shown in the quantifiers separately from the matrix *y succeeded x*.

The matrix of a statement always contains at least one pronoun. The second feature of the transformation to which we draw attention is the use of letters *x*, *y* as pronouns. We have used an example in which some such device is necessary to avoid ambiguity and it will be evident later that the same consideration explains the use of letters as pronouns in the case of some non-singular statements also. However, letters are used quite generally as pronouns in quantificational logic, even where there is no danger of ambiguity. This is mainly for the sake of consistency but there are certain additional advantages; for example we are not bothered by considerations of gender or inflexion. As applied to singular quantification this

[1] The reader should be warned, however, that the singular quantifier is not a standard element of quantificational logic. See chapter 2, section 8, first paragraph.

general use of letters as pronouns means for example that in expressing in quantificational form a proposition such as:

(8) Claudius became emperor in A.D. 41,

instead of using the form:

(9) Claudius he is such that he became emperor in A.D. 41,

we use:

(10) Claudius *x* is such that *x* became emperor in A.D. 41, which of course is shortened to:

(11) (Claudius *x*) *x* became emperor in A.D. 41.

It is usual to employ for this purpose letters from the end of the alphabet but it is hardly necessary to say that which particular letter or set of letters we use does not in general make any difference. Thus:

(12) (Claudius *y*) *y* became emperor in A.D. 41

has exactly the same meaning as (11), and

(13) (Claudius *z*) (Nero *y*) *y* succeeded *z*

has exactly the same meaning as (7′). A pronoun letter in the matrix refers to the noun which immediately precedes that letter in the quantifier and what particular letter is used does not matter so long as ambiguity is avoided. Letters used in this way as pronouns will be referred to as *pronoun variables* or sometimes just as *variables*.

4. An alternative formulation. The singular quantifiers (*Claudius x*), (*Nero y*) have hitherto been explained in such a way that (*Nero y*) is to be interpreted as meaning *Nero y is such that* which is a variant for *Nero he is such that*. *Nero he is such that* is equivalent to *Nero is such that* and this expression we defined as having the same meaning as the prefix *Of Nero it is true that*. It is not difficult to see that it would have been possible for us to evolve the singular quantifier (*Nero y*) in a more direct way from the prefix *Of Nero it is true that* without mentioning the *is such* *that* formula. Let us now do this. We find that the statement (1) *Nero succeeded Claudius* may be expressed in turn in the forms:

(2) Of Nero it is true that he succeeded Claudius,

(3*) Of Claudius it is true that of Nero it is true that he succeeded him.

However, (3*) is ambiguous and the ambiguity must be removed. A suitable device is to insert the phrase *as* x after *Claudius* and the phrase *as* y after *Nero* and replace the pronouns in the appropriate way by x and y. This gives us:

(6*) Of Claudius as x it is true that of Nero as y it is true that y succeeded x.

(6*) may be abbreviated to:

(7′) (Claudius x) (Nero y) y succeeded x.

(6*) is equivalent in meaning to (6′). (7) may be regarded as an abbreviation of either (6′) or (6*) and in general a singular quantifier of the form (a**x**) where a is a proper name may be regarded as an abbreviation for either of two expressions: (i) *a* **x** *is such that*, (ii) *Of a as* **x** *it is true that.*[1]

These two expressions are to be regarded as identical in meaning. They are, however, syntactically different and, since sometimes one and sometimes the other makes possible an easier explanation of an aspect of quantification, it seems advisable to keep both of them in mind.

5. Universes of discourse. Before going on to explain the quantification of general statements we must say something about the notion of a universe of discourse. Statements using the words *every*, *some* or their equivalents are fully significant only in relation to a set or group of indi-

[1] For an explanation of the use here of bold-faced letters see page 33.

viduals, usually a set of individual things or persons. Consider the statement:

(1) Everyone has learnt arithmetic.

This may be true or false according to the context in which it is made. If it is made with reference to the set of all individual persons in the world it is false; but if it is made with reference, for example, to the set of all university graduates then it is true. Again take the statement:

(2) There are some gold coins.

If this is made with reference to the set of all individual objects in the universe it is true, but it is false with reference to the contents of my pocket.

In a given context a set or class of individuals by reference to which the words *every* and *some* are to be understood in that context is a ***universe of discourse*** for that context; the term ***domain*** is frequently used in a similar sense. We have mentioned above four different universes of discourse: the set of individual persons in the world, the set of university graduates, the set of individual objects in the universe, the set of things contained in my pocket.

The elements of a universe of discourse may all be individual things or persons. Alternatively they may be properties of individuals or properties of properties; or some of them may be individuals and some of them properties. The system of logic expounded in this book is intended primarily to deal with arguments which presuppose universes of discourse consisting of individuals only. It is not confined to arguments of this kind but we do not attempt to define here the conditions under which it may be applied to arguments presupposing universes of discourse of other kinds.

A set which is used as a universe of discourse may contain either a finite or an infinite number of members. We assume throughout this book, however, that an empty set cannot be used as a universe of discourse, i.e. that any universe of discourse contains at least one element. Many writers indeed use the term *universe of discourse* in a sense such that an empty set also may be a universe of discourse; and the term *universe of discourse* as used by us is accordingly equivalent to *non-empty universe of discourse* in their terminology.

6. Quantifier-matrix form of general statements. We have seen that a singular statement can be expressed in the form of a statement which consists of a quantifier followed by a matrix. Such a statement is in *quantifier-matrix* form. General statements also may be expressed in quantifier-matrix form, though, naturally, different sorts of quantifier are needed; and we now go on to show how this may be done.

It is necessary to distinguish between qualified and unqualified general statements, i.e. between statements in which the generalizer is qualified in some way in relation to the universe of discourse and those in which it is not. We give some examples; for numbers (1) to (6) we assume a set of *things* as the universe of discourse and for numbers (7) to (12) a set of *men*.

Unqualified general statements	*Qualified general statements*
(1) Everything is fragile	(2) Everything made of glass is fragile
(3) The intruders damaged nothing	(4) The intruders damaged nothing valuable
(5) Something has been removed	(6) Something heavy has been removed
(7) Everyone blames the driver	(8) Everyone who does not know the facts blames the driver
(9) No one is happy	(10) No stranger is happy
(11) The evidence was destroyed by someone	(12) The evidence was destroyed by some well-meaning friend

1.6. *Quantifier-matrix*

An unqualified general statement can always be expressed in the form of a statement beginning with one or other of the two following expressions: ***Every individual is such that, Some individual is such that.*** The word ***individual*** here may be regarded as short for either ***individual thing*** or ***individual person*** according to the context. Thus (1) may be expressed as:

(1a) Every individual is such that it is fragile.

(1*a*) in turn may be expanded into (1*b*) which itself turns into (1*c*) when the pronoun *it* is replaced by the pronoun variable x.

(1b) Every individual it is such that it is fragile.

(1c) Every individual x is such that x is fragile.

In (1c) we have (1) expressed as a statement consisting of a quantifier, *Every individual x is such that*, followed by a matrix, *x is fragile*. The quantifier here is a *universal* quantifier and we shall use for it the symbol $(\forall x)$. Thus instead of (1) we may now write:

(1′) $(\forall x)$ x is fragile.

Now let us consider (11). (11) may be reduced to quantifier-matrix form by the following stages:

(11) The evidence was destroyed by someone.

(11a) Some individual is such that the evidence was destroyed by him.

(11b) Some individual he is such that the evidence was destroyed by him.

(11c) Some individual x is such that the evidence was destroyed by x.

In (11c) we have (11) expressed as a statement consisting of a quantifier *Some individual x is such that* followed by a matrix *the evidence was destroyed by x*. The quantifier in this case is known as a *particular* or *existential* quantifier. We symbolize it by $(\exists x)$ and with this notation may write for (11):

(11′) $(\exists x)$ the evidence was destroyed by x.

We have here explained the universal and particular quantifier symbols $(\forall x)$ and $(\exists x)$[1] as standing respectively for *Every individual x is such that* and *Some individual x is such that*. However, in the same way that the singular quantifier (ax) may be regarded as standing for either of the two expressions *a x is such that*, *Of a as x it is true that* which are equivalent to one another, so we may regard *Of every individual as x it is true that*, and *Of some individual as x it is true that* as variant forms of the universal and the particular quantifier respectively. In short the following equations hold:

(i) $(a\mathbf{x})=a$ **x** is such that = Of a as **x** it is true that.

(ii) $(\forall\mathbf{x})$ = Every individual **x** is such that = Of every individual as **x** it is true that.

(iii) $(\exists\mathbf{x})$ = Some individual **x** is such that = Of some individual as **x** it is true that.

In (i) a represents any singular term and in each equation **x** represents any pronoun variable, e.g. x, y, z.

We now give the quantifier-matrix form of the remaining unqualified general statements used as examples on page 11. (3) is equivalent to:

(3a) Everything is such that the intruders did not damage it,

which can be seen to be reducible to the Q-M statement:

(3′) $(\forall x)$ the intruders did not damage x.

(5), (6), (7) and (9) become respectively:

[1] $(\forall x)$ is sometimes read as 'For all x' or 'All x' and $(\exists x)$ as 'There exists an x' or 'Exists x'. (x) is often used instead of $(\forall x)$.

(5′) $(\exists x)$ x has been removed.

(7′) $(\forall x)$ x blames the driver.

(9′) $(\forall x)$ x is not happy.

We now come to qualified general statements. To express a qualified statement in quantifier-matrix form we first transform it into an unqualified statement of equivalent meaning; we do this by transforming it into a statement beginning with whichever is appropriate of the expressions: *Everything is such that*, *Everyone is such that*, *Something is such that*, *Someone is such that*. The reduction to quantifier-matrix form is then straightforward. As examples let us reduce (4) and (6) to quantifier-matrix form. The necessary steps are set out below:

(4) The intruders damaged nothing valuable.
Everything is such that if it is valuable the intruders did not damage it.
Every individual is such that if it is valuable the intruders did not damage it.
Every individual it is such that if it is valuable the intruders did not damage it.
Every individual x is such that if x is valuable the intruders did not damage x.

(4′) $(\forall x)$ if x is valuable the intruders did not damage x.

(6) Something heavy has been removed.
Something is such that both it is heavy and it has been removed.
Some individual is such that both it is heavy and it has been removed.
Some individual it is such that both it is heavy and it has been removed.
Some individual x is such that both x is heavy and x has been removed.

(6′) $(\exists x)$ both x is heavy and x has been removed.

In a qualified general statement the qualification appears to be a qualification of the generalizer: compare, e.g., *Something has been removed* with *Something heavy has been removed.* We see, in the two examples we have given, that on reduction of a qualified statement to quantifier-matrix form the qualification in a sense becomes not, as we might perhaps have expected, part of the quantifier but instead part of the matrix; a qualification of a universal generalizer giving rise to an *if–then* clause and that of a particular generalizer to a *both–and* clause in the matrix.[1] We give now the quantifier-matrix form of (2), (8), (10) and (12); a step-by-step reduction is shown for (2) and (12).

(2) Everything made of glass is fragile.
Everything is such that if it is made of glass it is fragile.
Every individual it is such that if it is made of glass it is fragile.
Every individual x is such that if x is made of glass x is fragile.

(2′) $(\forall x)$ if x is made of glass x is fragile.

(12) The evidence was destroyed by some well-meaning friend.
Someone is such that both he was a well-meaning friend and the evidence was destroyed by him.
Some individual he is such that both he was a well-meaning friend and the evidence was destroyed by him.
Some individual x is such that both x was a well-well-meaning friend and the evidence was destroyed by x.

(12′) $(\exists x)$ both x was a well-meaning friend and the evidence was destroyed by x.

[1] See, further, remarks on page 74 about formulae of the form $(\exists x)[P \supset Q]$.

(8) Everyone who does not know the facts blames the driver.

(8′) (∀*y*) if *y* does not know the facts *y* blames the driver.

(10) No stranger is happy.

(10′) (∀*x*) if *x* is a stranger *x* is not happy.

In each of these examples some pronoun variable other than the one actually used would do equally well.

We have now given the quantifier-matrix form of each of our twelve examples. The quantificational form of a statement may or may not be identical with its quantifier-matrix form, but before we discuss, as we will in section 9, the quantificational forms of (1) to (12) it will be convenient to explain the concept of a predicate and to deal with certain questions related thereto.

7. Predicate and argument[1]**.** By *singular statement* we shall mean any statement which contains one or more proper names, and by *unquantified singular statement* we shall mean a singular statement which contains no singular quantifier. Every unquantified singular statement may be regarded as the union of a *predicate* and one or more proper names. Consider the unquantified singular statement (i) *Socrates is wise*. This may be regarded as the union of the predicate ( ) *is wise* with the proper name *Socrates*, the union being formed when the name is inserted in the bracketed space within the predicate, and the brackets are removed. Consider again (ii) *Edinburgh is north of London*. We may regard this statement as having for predicate the formula (*1*) *is north of* (*2*); the predicate is united with the proper names *Edinburgh* and *London* when these names are

[1] It will be obvious that the word 'argument' is used here in an entirely different sense from that which it has borne in earlier sections.

inserted in place of *1* and *2* respectively within the predicate, and the two pairs of brackets are removed.

The bracketed space in the predicate of (i) and the numbered bracketed spaces in the predicate of (ii) will be referred to as *argument places* or just as *places*. Whatever term is put within such a place will be referred to as an *argument* of the predicate. Numbers are used as follows: when several argument places within a certain predicate are not numbered at all or are all numbered similarly this indicates that this is a predicate such that in any statement or other formula in which it occurs these places must all be filled by identical arguments; on the other hand if two places are numbered differently in a certain predicate this indicates that these places need not, though they may, be filled by identical arguments. A predicate is known as an *n*-place (or *n*-adic) predicate if the number of distinct arguments which it may take, i.e. which may occupy its argument places, is *n*. Thus the predicates in (i) and (ii) are 1- and 2-place predicates respectively; ( ) *is* ( )*'s worst enemy* is a 1-place predicate; (*1*) *knows better than* (2) *what* (2) *is doing* is a 2-place (or dyadic) predicate; (*1*) *is nearer to* (*3*) *than to* (2) is a 3-place predicate.

When all the argument places in a predicate are occupied by proper names we have a singular statement. One or more of them may be occupied by proper name variables a, a_1, ... , and the remainder, if any, by proper names, and in this case we have a statement form; e.g., *a is wise* is a statement form in which the predicate ( ) *is wise* is united with the argument *a*. But it is possible for an argument-place in a predicate to be occupied not by a proper name or proper name variable but by a pronoun variable. This is so in the following formulae for example: *x is wise*, *y is north of London*, *x is greater than a*, *y is nearer to x than to z*. Formulae such as these are neither statements nor statement forms. They are, however, possible components of statements or statement forms; we shall refer to them as *pronominal clauses*. Thus the pronominal clause *x is wise* is a component of the statement (iii) $(\exists x)$[*x is a Greek and x is wise*], *x is greater than a* is a component of the statement form (iv) $(\forall x)$ [*x is greater than a*] and *y is north of London* is a component of the statement (v) (*Edinburgh y*)[*y is north of London*]. We see that a pronominal clause may be the matrix of a quantified statement or statement form; for example *y is north of London* is the matrix of (v) and *x is greater than a* is the matrix of (iv). On the other hand it may be a component of the matrix of a quantified statement or statement form, though not itself identical with the matrix. It is indeed true that when a pronominal clause occurs in a statement it is always as the matrix or a component of the matrix of a quantified statement.

8. Predicate abbreviation and substitution. The evaluation of quantificational arguments can be greatly facilitated by the use of abbreviations. Abbreviations for three different types of expression may be used: for proper names, for whole statements which do not require analysis and for predicates. In the first two cases the processes of abbreviating and interpreting abbreviations are straightforward but in the case of predicates some explanation may be helpful. As an abbreviation for a predicate we use a capital letter (as a rule other than *F*, *G* or *H*) followed by the appropriate number of argument places which, if the number is greater than 1, are numbered in order from left to right: (*1*), (*2*), (*3*), ... Thus for the predicate (   ) *is wise* we might use the abbreviation *W*( ), for (*1*) *is north of* (*2*), we might use *N*(*1*)(*2*) and for (*1*) *is nearer to* (*3*) *than to* (*2*) we might use *R*(*1*)(*2*)(*3*). A predicate that occurs as part of a statement or statement form has as argument in each argument place either a proper name (e.g. *Socrates*), a proper name variable (e.g.

a, a_1), a pronoun variable (e.g. *x*, *y*, *z*) or, as we shall see later, a dummy name (e.g. *d*, d_1). For example, in the statement *Socrates is wise* the predicate ( ) *is wise* has as argument the proper name *Socrates*; in the formula *x is wise* it has as argument the pronoun variable *x*. Again in the formula *Edinburgh is nearer to x than to a* the predicate (*1*) *is nearer to* (*3*) *than to* (*2*) has in the argument places numbered (*1*), (*2*), (*3*) respectively the proper name *Edinburgh*, the proper name variable *a* and the pronoun variable *x*. To replace a predicate at a given occurrence in a statement or statement form by an abbreviation we must first have a key; this can conveniently be set out in two columns, with predicate on the right and abbreviation on the left. In the predicate column immediately below the predicate we write the formula which is to be abbreviated. Then on the left we copy down the capital letter and under each argument place the corresponding term from the right-hand column, following the numbers if there are more than one, so that no term appears under any number on one side without also appearing under that number on the other side. The resulting entry in the left-hand column is the required abbreviation. For example if we want to abbreviate the predicate (*A*) (*1*) *is nearer to* (*3*) *than to* (*2*) in the context of the formula (*A'*) *Edinburgh is nearer to x than to a*, we first write down a key giving the proposed predicate abbreviate, e.g.

ABBREVIATION	PREDICATE
R (1) (2) (3)	(1) is nearer to (3) than to (2)

Now we write the formula (*A'*) immediately under the predicate: this shows what argument places the arguments *Edinburgh*, *x* and *a* must occupy in the abbreviation and so we are able to fill in the second line of the left-nand column which gives us the abbreviated formula we require. The completed table is:

R (1) (2)(3)	(1) is nearer to (3) than to (2)
R Edinburgh *a* *x*	Edinburgh is nearer to *x* than to *a*

If we have a key set out as before in two columns and a formula abbreviated in accordance with the key, the procedure described can be used in the opposite direction to give us the formula in full, e.g., given the top line in the following table as key, by writing the abbreviated formula *Kyxx* immediately under the left-hand entry we obtain the information which enables us to complete the right-hand column and so obtain the full formula of which *Kyxx* is an abbreviation.

K(1)(2)(3)	(1) knows better than (2) what (3) is doing
K y x x	*y* knows better than *x* what *x* is doing

In the same way it can be seen that with the same key the formula *Kz Smith y* is an abbreviation for *z knows better than Smith what y is doing*.

Essentially the same procedure which we have just been describing, i.e. the replacement in a formula of a predicate abbreviation by a full predicate, can be used in another operation concerned with predicates namely the substitution in a formula for a predicate variable of a predicate. As we shall see, a quantificational statement form may contain, as well as variables of other kinds, a number of predicate variables, and in order to obtain an exemplification of such a form we have to substitute a predicate for each predicate variable. As an *n*-place predicate variable we use one of the capital letters *F*, *G*, *H* followed by *n* argument places, numbered in order from left to right if the number is greater than one. A predicate variable is united with arguments in the same way that a predicate is united with arguments and, when it occurs in a formula, a predicate variable has in each argument place a proper name, a proper name variable, a pronoun variable or a dummy name. To substitute a predicate *P* for a predicate variable in a formula at a given occurrence it is sufficient to treat the predicate variable as if it were an abbreviation for P and follow the procedure described in the last paragraph for the replacement of a predicate abbreviation by a full predicate. For example, if in the following formula:

(i) $(\forall x)[Fx \supset (\exists y)[Fy \,.\, Gyx \,.\, \sim Gxy]]$,

we substitute for the predicate variable $F(\ )$ the predicate

( ) *is a number*

and for the predicate variable $G(1)(2)$ the predicate

(*1*) *is greater than* (*2*),

we obtain the following statement which is an exemplification of (i):

(i) $(\forall x)[x$ is a number $\supset (\exists y)[y$ is a number . y is greater than x . $\sim x$ is greater than $y]]$.

9. Quantificational form. Strictly, a statement which is premiss or conclusion of an argument has quantificational form only in relation to the argument. Consider the statements (1) to (12) which we use to illustrate the notion of quantifier-matrix form in section 6. These statements were written down in isolation, not in the context of arguments. Their quantifier-matrix form is independent of context but the same is not true of their quantificational form. For any one of these statements different quantificational forms might be appropriate according to the argument in which it occurred, and even with one definite argument as context different quantification forms could still be appropriate inasmuch as a number of different analyses of the argument into elements might be possible.

The notion of an *analysis of an argument into elements* is not fundamentally a difficult one and can probably be grasped most easily by a study of examples. An element of an argument is either a statement or a predicate, and broadly a statement or a predicate can be regarded as an element of an argument if for the purpose of the argument it can be treated as a unit. More strictly we may say that in any one analysis of an argument into elements a statement or predicate F may be treated as an element if but only if

(i) it occurs within the argument;

and

(ii) no statement or predicate which is a component of F either (a) occurs in the argument otherwise than as part of F, or (b) has been treated earlier in the analysis as an element.

Consider the following arguments: (A) $(\forall x)$[*If x is a newcomer x has settled down and x is happy*]; *Smith is a newcomer; therefore Smith has settled down and Smith is happy*. (B) $(\forall x)$[*If x is a newcomer x has settled down and x is happy*]; *Smith is a newcomer; therefore Smith is happy*. The statement *Smith is a newcomer* cannot be treated as an element of (A) because it has as a component the predicate (   ) *is a newcomer* which occurs elsewhere in the argument as a component not of this statement but of the pronominal clause *x is a newcomer*. However, the predicate (   ) *is a newcomer* may be treated as an element of (A), and likewise of (B). The predicate (   ) *has settled down and* (   ) *is happy* which occurs in the premiss can be treated as an element of (A) provided that (   ) *has settled down* or (   ) *is happy* has not already been treated as an element since, although the two components do occur elsewhere in the argument, they do so only as parts of this predicate. On the other hand this same predicate cannot be treated as an element of (B) since in that argument the component (   ) *is happy* occurs in isolation in the conclusion. There is only one possible analysis of (B) namely into the three elements: ( ) *is a newcomer*, ( ) *has settled down* and ( ) *is happy*. There are two possibilities for (A): it may be regarded as having the same elements as B but alternatively it may be treated as having the two elements: ( ) *is a newcomer* and ( ) *has settled down and* ( ) *is happy*. Consider again (C) *If* $(\exists x)[x$ *escapes*$]$ $(\forall x)[x$ *will rejoice*$]$; *therefore if* $(\exists x)[x$ *escapes*$]$ *Jones will rejoice*; and (D) *If* $(\exists x)[x$ *escapes*$]$ $(\forall x)[x$ *will rejoice*$]$; *therefore if Smith escapes Jones will rejoice*. As elements of (D) we must take the two predicates ( ) *escapes* and ( ) *will rejoice*. But as elements of (C) we may take either these same two predicates or the predicate ( ) *will rejoice* and the statement $(\exists x)[x$ *escapes*$]$. If we can choose between a predicate and a statement for element in a given analysis we should choose the statement: the argument will be more easily handled; indeed if instead of (C) we had the argument (C′) which is the same as (C) except that $(\exists x)[x$ *escapes*$]$ is replaced at both occurrences by the equivalent statement *someone escapes* treatment of this statement as an element will obviate any necessity for expressing it quantificationally at all. Again, provided our analysis is complete the smaller the number of elements the better. Thus for both (A) and (C) the second of the two possible analyses is the better one.

We now define quantificational form as follows:

> A statement, S, is in quantificational form with respect to an argument of which it is a premiss or the conclusion if and only if, for a given analysis of that argument into elements, no element which is implicit in S occurs in S otherwise than as a truth-functional component of S or of a matrix or as a predicate component of a truth-functional component of S or of a truth-functional component of a matrix.

It is to be understood that for the purposes of this definition a formula is regarded as a truth-functional component of itself.

To illustrate the application of this definition let us consider again the twelve statements which were used as examples in section 6. The quantifier-matrix form of these statements has been given as:

(1′) $(\forall x)$ [x is fragile]

(2′) $(\forall z)$ [if z is made of glass z is fragile]

(3′) $(\forall x)$[the intruders did not damage x]

(4′) $(\forall x)$ [if x is valuable, the intruders did not damage x]

(5′) $(\exists x)$ [x has been removed]

(6′) $(\exists x)$ [both x is heavy and x has been removed]

(7′) $(\forall x)$ [x blames the driver]

(8′) $(\forall y)$ [if y does not know the facts y blames the driver]

(9′) $(\forall x)$ [x is not happy]

(10′) $(\forall x)$ [if x is a stranger x is not happy]

(11′) $(\exists x)$[the evidence was destroyed by x]

(12′) $(\exists x)$[both x was a well-meaning friend and the evidence was destroyed by x]

The examples (1) to (12) have been given without context. Let us suppose arbitrarily that (1) occurs in the context of an argument in an analysis of which (   ) *is fragile* is taken as one of the elements; in this case we see that (1) is already in quantificational form (as 1′) with respect to that argument since the only element (i.e. (   ) *is fragile*) which occurs in it occurs as a matrix. Let us similarly make arbitrary suppositions about the other examples, to the effect that the elements occurring in (2) to (11) are as follows: (2) (   ) *is made of glass*, (   ) *is fragile*; (3), (4) *the intruders damaged* (   ), (   ) *is valuable*; (5) (   ) *has been removed*; (6) *both* (   ) *is heavy and* (   ) *has been removed*; (7), (8) (   ) *blames the driver*, (   ) *knows the facts*; (9), (10) (   ) *is not happy*, (   ) *is a stranger*; (11), (12) *the evidence was destroyed by* (   ), (   ) *was a well-meaning friend*. On these suppositions it can be seen that (5), (6), (7), (9) and (11) are already in quantificational form as (5′) (6′), (7′), (9′) and (11′) respectively since in each case the only element which occurs is the predicate component of a matrix. (3′) and (4′) are not in quantificational form since an element which is implicit in both, *the intruders damaged* (   ), which is a predicate, does not occur in either case as predicate component of a truth-functional component either of the statement itself or of a matrix. However, if in both cases we replace *the intruders did not damage x* by ~ *the intruders damaged x* and in (4′) also replace *if* in the appropriate way by ⊃ we obtain:

(3*) $(\forall x)$[~ the intruders damaged x]

and

(4*) $(\forall x)$[x is valuable ⊃ ~ the intruders damaged x].

(3*) and (4*) are now in quantificational form. In (3*) the only element, *the intruders damaged* (   ), occurs as the predicate component of a truth-functional component of a matrix; in (4*) this predicate and the only other element (   ) *is valuable* occur in each case as predicate components of a truth-functional component of a matrix. The quantificational forms of the remaining statements can now be seen to be:

(2*) $(\forall z)$[z is made of glass ⊃ z is fragile];

(8*) $(\forall y)$[~ y knows the facts ⊃ y blames the driver];

(10*) $(\forall x)$[x is a stranger ⊃ x is not happy];

(12*) $(\exists x)$[x was a well-meaning friend . the evidence was destroyed by x].

On the stated supposition about context the statements (1) to (12) have now been expressed in quantificational form. The rules of quantificational logic could be applied to them as they are. However, they can be dealt with more easily if abbreviations are introduced for the elements. The following scheme of abbreviation might be adopted:

$F(\ )$=( ) is fragile
$D(\ )$=the intruders damaged ( )
$R(\ )$=( ) has been removed
$B(\ )$=( ) blames the driver
$P(\ )$=( ) is not happy
$T(\ )$=the evidence was destroyed by ( )
$G(\ )$=( ) is made of glass
$V(\ )$=( ) is valuable
$Q(\ )$=both ( ) is heavy and ( ) has been removed
$K(\ )$=( ) knows the facts
$S(\ )$=( ) is a stranger
$W(\ )$=( ) was a well-meaning friend

With abbreviations in accordance with this scheme the twelve statements become:

(1**) $(\forall x)Fx$ (2**) $(\forall z)[Gz \supset Fz]$
(3**) $(\forall x) \sim Dx$ (4**) $(\forall x)[Vx \supset \sim Dx]$
(5**) $(\exists x)Rx$ (6**) $(\exists x)Qx$
(7**) $(\forall x)Bx$ (8**) $(\forall y)[\sim Ky \supset By]$
(9**) $(\forall x)Px$ (10**) $(\forall x)[Sx \supset Px]$
(11**) $(\exists x)Tx$ (12**) $(\exists x)[Wx \,.\, Tx]$

By our supposition the elements in all cases have been predicates. However, if we had supposed that a statement was itself an element we should have regarded it as being in quantificational form from the beginning without need of transformation. As an abbreviation for a statement which is an element we need only a single letter. Thus if (1) *Everything is fragile* were an element it would be its own quantificational form and could be abbreviated to, for example, *e*.

We have been dealing with the expression in quantificational form of comparatively simple types of general statement, those in which there is a single generalizer only, which governs the whole statement. Later on we shall set forth a system of rules whereby any general statement of any degree of complexity may be reduced to quantificational form. However, there is a certain complication which may arise in connexion with the expression in quantificational form of even simple kinds of general argument. This we discuss in the next section.

10. The question of existential import. The traditional syllogistic logic recognized four main types of general statement known respectively by the letters *A*, *E*, *I*, *O*. The four types are: (*A*) *Every f is a g*, (*E*) *No f is a g*, (*I*) *Some f is a g* and (*O*) *Some f is not a g*. *A* and *E* are universal statements; *I* and *O* are particular. *A* and *I* are affirmative; *E* and *O* are negative. The first three types of statement are represented among our examples of qualified general statements in the last section. (2) and (8) can be expressed as *A*-statements, (4) and (10) as *E*-statements. (6) and (12) are *I*-statements. It can be seen from our treatment of these examples that (if we put Fx for *x is an f*, and Gx for *x is a g*) the four types can be expressed quantificationally as follows, the letters *a*, *e*, *i* and *o* respectively being used to mark the quantificational expression of *A*, *E*, *I*, and *O* statements respectively: (*a*) $(\forall x)[Fx \supset Gx]$; (*e*) $(\forall x)[Fx \supset \sim Gx]$; (*i*) $(\exists x)[Fx \,.\, Gx]$; (*o*) $(\exists x)[Fx \,.\, \sim Gx]$. The correctness of these renderings for the universal types of statement *A* and *E* may be questioned on the ground that the *A* and *E* statements have, whereas the *a* and *e* statements do not have, existential import with respect to *f*'s. This means, to confine the discussion to the case of *A* and *a*, that the *A*-statement *Every f is a g* cannot be true unless at least one *f* exists whereas the *a*-statement $(\forall x)[Fx \supset Gx]$ ($=(\forall x)$[*if x is an f x is a g*]) can be true even if there are no *f*'s. In view of this difference it is suggested the *a*-form cannot be a correct rendering of the *A*-form. Now it is certainly true that the *a*-statement does not have existential import: $(\forall x)[Fx \supset Gx]$ not only can but actually must be true if there are no *f*'s.[1] Further, with regard to

[1] If there are no *f*'s, Fx is false of every individual as x; hence, in accordance with the truth table for $\supset$, $Fx \supset Gx$ must be true of every individual as x.

the *A*-statement we must allow that at least in very many contexts the person who makes the statement believes that *f*'s exist. Let us suppose that this is true of all contexts in which the assertion *Every f is a g* is made: does it follow that we ought to regard the *A*-statement as equivalent not to the *a*-statement but to the conjunction of the *a*-statement with a statement asserting the existence of at least one *f*, i.e. to the compound statement:

$$(\forall x)[Fx \supset Gx] \,.\, (\exists x)Fx?$$

It is important to realize that this does not follow. We must remember that, as well as contexts in which the *A*-statement is made as a complete assertion, i.e. in which the simple assertion is made that every *f* is a *g*, there are other contexts in which it occurs as a component of a compound statement; in particular there are cases in which it is denied to be true. Let us consider as a particular example the statement (A_1) *Every fair-haired competitor is a German.* It is certainly true that in most, if not all, contexts in which we can imagine this statement being made on its own, as a genuine assertion, the person who asserted it would have the belief that there is at least one fair-haired competitor. But equally it is true of the following statement:

(A_1') It is not the case that every fair-haired competitor is a German,

which contains A_1 as a component that in most contexts the person asserting it would be likely to have the belief that there was at least one fair-haired competitor. If what we mean by saying that a statement has existential import with respect to *f*'s is that the person who makes the statement believes that *f*'s exist then we may say that both (A_1) and its negation (A_1') are likely to have existential import. However, if an *A*-statement were regarded as simply equivalent to the conjunction of the *a*-statement and a statement asserting the existence of *f*'s then its negation would not have existential import; for its negation would be equivalent to the negation of the conjunction in question, i.e. to:

It is not the case that $[(\forall x)[Fx \supset Gx] \,.\, (\exists x)Fx]$,

and this negative statement could certainly be true if there were no *f*'s and could be believed to be true by someone who believed that there were no *f*'s. For this reason it would not be satisfactory to regard the *A*-statement as equivalent in meaning to the conjunction in question.[1]

The view which we take in this book is that the *A*-statement does not have the existence of *f*'s as part of its meaning: its meaning is satisfactorily expressed by the *a*-statement. Nevertheless in many contexts a person making an assertion of which the *A*-statement is a component part or the whole does assume the existence of *f*'s and perhaps would not think it appropriate to make the assertion in the absence of this assumption. In such a case in order to do justice to his thought when we express it in quantificational form it may be necessary to give separate expression to this assumption. To illustrate this we may point out that while the argument, $(\forall x)[Fx \supset Gx]$; *therefore* $(\exists x)[Fx \,.\, Gx]$, is certainly invalid a person who says: *Every f is a g; therefore some f is a g*, is not necessarily, and indeed is probably not, reasoning fallaciously; for in drawing his conclusion he may be relying not only on the belief that $(\forall x)[Fx \supset Gx]$ (i.e. that everything is such that if it is an *f* it is a *g*) but also on the unexpressed belief that $(\exists x)Fx$ (i.e. that something is an *f*). Hence his argument may fairly be expressed in quantificational form as: $(\forall x)[Fx \supset Gx]$, $(\exists x)Fx$; *there-*

[1] A point substantially similar to this is made by P. F. Strawson in the course of an interesting discussion of existential import and related questions. See Strawson, *Introduction to Logical Theory*, chapter 6.

fore $(\exists x)[Fx \,.\, Gx]$, which is valid. The question of whether in a given case any such existential addition is necessary when an argument is being expressed quantificationally is one which can be decided only by a study of the context. In the last section the examples (2), (4), (8) and (10) have been given without context: we have shown how their meanings may be expressed quantificationally but in some contexts at least an additional existential statement would be necessary.

In short our view might be put thus: existential import with respect to f is something which does not belong to the A-statement or the E-statement as such, but it may belong to a given context in which the A- or the E-statement is made. The view underlying the traditional syllogistic logic, in its commonest form, appears to be that existential import with respect to f belongs to all contexts in which the A- or the E-statement occurs. However, existential import does not, on the traditional view, belong to the A- or the E-statement itself in the sense that existence is part of the meaning of these statements; for to assert that it does would be incompatible with traditional treatment of the O-statement as equivalent to the simple negation of the A-statement and of the I-statement as equivalent to the simple negation of the E-statement.

The traditional logic not only requires that existential import with respect to f should belong to all contexts in which A- or E-statements occur. It really requires that existential import with respect to any term f or g contained in a statement should belong to any context within which that statement occurs: it is a system of logic which is intended to be applied to statements of the A, E, I or O types in which the general terms f and g designate only non-empty classes. Given this condition the traditional square of opposition shows correctly the interrelationships between the four types of statement:

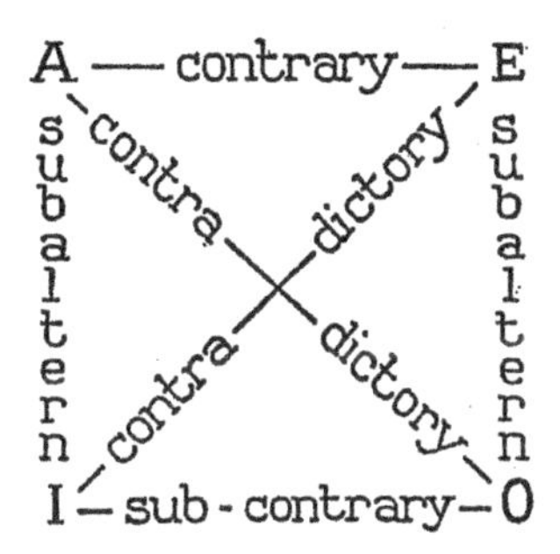

Contradictory statements, e.g. *A* and *O* can be neither both true nor both false; contrary statements *A*, *E* can both be false but cannot both be true; sub-contrary statements *I*, *O* can both be true but cannot both be false; the subaltern to a true statement must itself be true. However, if we consider the *a*, *e*, *i* and *o* statements on the other hand, and make no stipulations about contexts, we see that the corresponding relationships hold only for the pairs *a*-*o* and *e*-*i*. *a* and *o* can neither be both true nor both false, and the same applies to *e* and *i*. But on the other hand if there are no *f*'s the *a*-statement $(\forall x)[Fx \supset Gx]$ and the *e*-statement $(\forall x)[Fx \supset \sim Gx]$ will both be true; the *i*-statement and the *o*-statement will both be false and the relations corresponding to those between *A* and *I* and *E* and *O* respectively will not hold; for the *a*-statement will be true but the *i*-statement false and the *e*-statement true but the *o*-statement false.

11. Definitions. In this section we enumerate some of the main kinds of expression and symbol which are used in the remainder of the book and define some important terms. We begin by listing the names of various kinds of expression or symbol and showing examples to the right.

Singular quantifiers:	(Socrates x), (Napoleon y), (ax), (a_1x), (dx)
Particular quantifiers (= existential quantifiers):	$(\exists x)$, $(\exists y)$
Universal quantifiers:	$(\forall x)$, $(\forall z)$
Proper names:	Socrates, Napoleon, Paris
Proper name variables:	a, a_1, a_2
Dummy names:	d, d_1, d_2
Pronoun variables:	x, y, z, x_1
Predicates:	( ) is wise, ( ) runs, (1) is north of (2)
Predicate variables:	$F(\ )$, $G(\ )$, $H(\ )$, $F(1)(2)$, $G(1)(2)$, $F(1)(2) \ldots (n)$
Statements:	Socrates is wise, Edinburgh is north of London, $(\exists x)$[x is wise], $(\exists x)(\exists y)$[x is north of y]
Statement variables (= propositional variables):	p, q, r
Statement forms:	p, $p \supset q$, F Socrates, a is wise, Fa_1, $(\exists x)Fx$, $p \supset (\forall y)$ [G Socrates y]
Pronominal clauses:	x is wise, x is north of y, $(\exists x)$[x is north of y]

A singular term is any symbol which is either a proper name or a dummy name.

An individual symbol is any symbol which is either a singular term or a pronoun variable or a proper name variable.

Our use of distinct symbols to serve as proper name variables, dummy names and pronoun variables respectively is not orthodox. The letters x, y, z are frequently used in all three roles.

Bold-faced type will sometimes be used. **a** represents any proper name or proper name variable, e.g., a, a_1, a_2, ... **d** represents any dummy name, e.g., d, d_1, d_2, ... **x** represents any pronoun variable, e.g., x, y, z. The Greek letter α is sometimes used to represent any singular term. F, P, Q are sometimes used to represent any formulae. The symbols mentioned in this paragraph are often used subject to certain restrictions which are stated in the contexts.

The symbols a, a_1, a_2, ... and the symbols which are classed above as predicate variables will sometimes be used not as variables but *ad hoc* as abbreviations for stated proper names or predicates: e.g., a for *Socrates*, $F(\ )$ for (   ) *is wise*.

Governing, scope. When a quantifier occurs in a statement it must have associated with it a certain matrix which immediately follows it. The matrix associated with a certain quantifier is said to be the *scope* of that quantifier and to be the matrix of the whole statement or clause which consists of quantifier followed by matrix; the quantifier is said to *govern* the whole statement or clause which consists of quantifier followed by matrix. For example, in

(1) $(\exists x)$ x was an officer. $(\forall y)$[y was an officer $\supset$ y was born in Great Britain]

the quantifier $(\forall y)$ has as its associated matrix the formula [*y was an officer* $\supset$ *y was born in Great Britain*]. This formula is the scope of that quantifier and the matrix of the whole clause: $(\forall y)$[*y was an officer* $\supset$ *y was born in Great Britain*]. $(\forall y)$ governs this whole clause.

Strictly, a matrix is bounded by a pair of square brackets which are to be regarded as forming part of it. However, where the scope of a quantifier is the shortest properly formed propositional expression (pronominal clause or statement) which immediately follows the quantifier the brackets may be omitted. Thus in (1) the scope of $(\exists x)$ is *x was an officer*; the full expression of (1) being $(\exists x)$ [*x was an officer*]. $(\forall y)$[*y was an officer* $\supset$ *y was born in Great Britain*]; similarly: $(\exists y)(\exists x)$ *x is north of y* may be written instead of $(\exists y)[(\exists x)$[*x is north of y*]].

Bound, free, captured. A quantifier is said to be an **x**-quantifier if it contains the variable **x**; e.g. $(\exists y)$ is a *y*-quantifier. An occurrence of a pronoun variable **x** is said to be *bound* in a formula *F* if it occurs in or within the scope of an **x**-quantifier which itself is within *F*. An occurrence of **x** is *free* in *F* if it is not bound in *F*. Thus the two occurrences of *x* and the three occurrences of *y* are bound in (1), but *y* is not bound in the formula [*y was an officer* $\supset$ *y was born in Great Britain*] at either occurrence. An individual symbol α is said to be *free for* **x** *in F*, if no occurrence of α in *F* lies within the scope of an **x**-quantifier. It follows that if α is not free for **x** in *F* at least one occurrence of α in *F* will be within the scope of an **x**-quantifier. In this case if α at this occurrence is replaced by **x**, **x** will be said to be *captured* by that **x**-quantifier; i.e. it will become bound by it. Examples are given in chapter two, section 5.

CHAPTER TWO

1. Quantificational validity. In the main part of this chapter we describe a method, the deductive method, which can be used to construct, or to demonstrate the validity of, valid quantificational arguments. In a later chapter we shall deal with proofs of invalidity. As a preliminary to these two tasks we must examine the concepts of validity and invalidity in their application to quantificational argument forms. Let us begin by repeating the definition of validity and invalidity for argument forms in general which was given in TFL.

> A form of argument is said to be *valid* if there is no possible exemplification in which the premisses are all true but the conclusion is false. It is invalid if in at least one possible exemplification all the premisses are true but the conclusion is false.[1]

Now a quantificational argument form may contain, apart from pronoun variables, variables of any or all of three kinds: propositional or statement variables p, q, r, ... , predicate variables F, G, H, ... , and proper name variables a, a_1, a_2, ... An exemplification of a quantificational argument form Q is obtained by making consistently substitutions of the appropriate kind for all the variables, other than pronoun variables, in Q, statements for the statement variables, predicates for the predicate variables and proper names for the proper name variables. Thus to take a simple example, where the form contains, apart from pronouns, a single predicate variable only, we obtain

[1] TFL, page 8.

an exemplification of the argument form, (A) ($\exists x$)Fx; *therefore* ($\forall x$)Fx, by substituting[1] for F( ) the predicate (   ) *was Prime Minister of the U.K. in* 1961. This gives us (A_1) ($\exists x$) [x *was Prime Minister of the United Kingdom in* 1961]; *therefore* ($\forall x$) [x *was Prime Minister of the United Kingdom in* 1961]. If we know that the premiss of A_1 is true and the conclusion false then of course we may say at once that A is invalid in accordance with our definition. However, as it happens, we are not yet able to say, without taking anything for granted, whether or not A_1 has premiss true and conclusion false. The reason for this is not lack of historical information but rather absence of information about the universe of discourse with respect to which the quantifiers ($\exists x$) and ($\forall x$) are to be interpreted. If the intended universe of discourse is a class consisting of Mr. Macmillan and a number of other persons, then of course A_1 has premiss true and conclusion false. But, on the other hand, if the universe of discourse is the class consisting of Mr. Macmillan alone or is a class to which Mr. Macmillan does not belong, then it is not the case that A_1 has premiss true and conclusion false. It is possible therefore for an exemplification of a quantificational argument form to have premisses true and conclusion false when construed with respect to one universe of discourse although it does not have such a property when construed with reference to some other universe. Because of this relativity our general definition of validity would not be entirely unequivocal when applied to quantificational arguments and we must now reformulate it to take account of their special features. Our new definition will be:

> A form of argument Q is valid if, with respect to every universe of discourse, there is no possible exemplification of Q in which all the premisses are true but the conclusion is false. It is invalid if with respect to at

[1] See chapter 1, section 8, last paragraph.

least one universe of discourse there is at least one possible exemplification of Q in which all the premisses are true but the conclusion is false.

Corresponding to this definition of validity for quantificational argument forms the following definition of logical truth for quantificational statement forms may also be given:

A quantificational statement form S is logically true if and only if every possible exemplification of S is true with respect to every universe of discourse.

It may be added that many writers apply the term 'valid' not only to argument forms but also to statement forms. When a statement form is said to be valid what is meant is that it is logically true in the sense just defined. We do not follow this usage here.

2. Deduction. We shall now describe a deductive method of constructing a valid argument or argument form. A rule of deduction is valid if it allows deduction in accordance with a valid argument form. It can be seen that an argument must be valid if its conclusion can be deduced from its premisses in a number of steps each one of which is in accordance with a valid rule and that an argument form must be valid if from any set of statements exemplifying its premisses the corresponding statement exemplifying its conclusion can be deduced in a number of steps each of which is in accordance with a valid rule. Hence if we can establish the validity of a number of simple argument forms, deductive rules based on these forms may be used to construct and to prove the validity of other valid arguments and argument forms. We shall now set forth a suitable set of valid rules and give examples of their application.

Any worthwhile set of rules for quantificational deduction must include as a part some set of rules of truth-functional deduction. The truth-functional rules which we shall use to begin with are those given in TFL, chapter four. These rules are incorporated, as (1) to (9) and (10) (equivalences I to XI), in the set of rules given in section 1 of chapter three of this book.[1]

In addition to truth-functional rules we have to have some rules which are not required in any purely truth-functional deduction but are peculiar to the non-truth-functional part of quantificational deduction. We will refer to these as quantificational rules. We give first a set of six starred rules: EI*, EG*, UI*, UG*, SQ* and SD*. The validity of EG*, UI*, SQ* and SD* will be evident as soon as these rules are understood. Some explanatory comment will be made on EI* and UG*.

We shall regard these starred rules as theoretically fundamental. Later, however, it will be shown that from them (together of course with the truth-functional rules) may be derived a set of four quantificational rules which are equivalent in their effect (subject to a stated qualification) to the six but are for practical purposes more convenient.

3. The rule EI*; dummy names. We begin with the rule EI*. This rule is that if we have a line *l* of a deduction which is governed by a particular quantifier (∃**x**) we may write as a new line the formula which is the same as *l* except that in place of the particular quantifier (∃**x**) it has the singular quantifier (**dx**), where **d** is a dummy name which has not previously occurred in the deduction. We may express this rule in abbreviated form thus:

EI* $(\exists \mathbf{x})P \rightarrow (\mathbf{dx})P$, where **d** is a dummy name not previously used.

A dummy name is a special symbol which we use tempor-

[1] See also, however, chapter 3, section 2.

arily in a deduction to refer to an individual whose identity (and whose real name) is undetermined. The use of dummy names in formal deduction has a familiar counterpart in ordinary discourse. It is quite common to make an indefinite statement about 'some' individual and then to stipulate that a certain particular name or symbol will be used to refer to this individual whoever he may be. For example, one might hear or read something like the following: 'Someone burgled this house last night. We don't know who it was but let us call him "Raffles" (or let us call him "*X*"). Raffles then burgled this house last night. He must have entered by the window because . . .' 'Raffles' or '*X*' is here used as a dummy name and the move from:

(1) Someone burgled this house last night,

to

(2) Raffles burgled this house last night,

corresponds to a move in accordance with EI*. In the light of this illustration the condition to which a move in accordance with EI* is subject, namely that the dummy name used must be one not previously used in the deduction becomes intelligible. Suppose that in addition to the statement (1) we have another statement about someone, e.g.

(3) Someone burgled this house two nights ago.

If we were to use the same dummy name *Raffles* to refer to this someone also, we should be committed to the assumption that the person who burgled the house two nights ago is the person who burgled it last night. This is an assumption which we are obviously not entitled to make purely on the basis of the two statements (1) and (3).

As dummy name symbols we shall use the letters $d, d_1, d_2, \ldots$. A dummy name is a species of singular term. If we express (1) in our example above in quantificational form as:

(1′) $(\exists x)[x$ burgled this house last night$]$

then by EI* using *d* instead of *Raffles* as a dummy name we obtain, corresponding to (2):

(2′) $(dx)[x$ burgled this house last night$]$.

4. The rules EG*, UI* and UG*. EI* is a rule which enables us to replace an initial particular quantifier by a singular quantifier. The other three rules in this group also have to do with the replacement of one type of quantifier by another. EG* is concerned with the replacement of a singular by a particular quantifier, UI* with the replacement of a universal by a singular quantifier and UG* with the replacement of a singular by a universal quantifier. EG* and UI* are quite general but certain conditions attach to UG*.

EG* then is the rule that if we have a line *l* of a deduction which is governed by a singular quantifier $(\alpha\mathbf{x})$, where α is any singular term, whether a proper name or a dummy name, we may write as a new line one which is the same as *l* except that in place of the singular quantifier $(\alpha\mathbf{x})$ it has the particular quantifier $(\exists\mathbf{x})$. Symbolically:

(EG*) $(\alpha\mathbf{x})P \rightarrow (\exists\mathbf{x})P$, where α is any singular term and $\mathbf{x}$ is any variable.

The validity of EG* is intuitively obvious provided that the meanings of the singular and particular quantifiers are understood. For if with respect to any universe of discourse it is true that a given individual α which belongs to the universe is such that P, then it must be true that some individual which belongs to the universe is such that P, whatever formula P may be.

UI* is the rule that if we have a line *l* of a deduction which is governed by a universal quantifier $(\forall\mathbf{x})$ we may write as a new line the formula which is the same as *l* except that $(\forall\mathbf{x})$ is replaced by $(\alpha\mathbf{x})$, where $\mathbf{x}$ is any variable and α is any singular term. Symbolically

$$(\forall \mathbf{x})P \rightarrow (\alpha\mathbf{x})P.$$

UI* needs no explanation. It is obvious that if for any universe of discourse it is true that every individual in the universe is such that P then it must be true that a given named individual α in that universe is such that P.

UG* is the rule that if we have a line l of a deduction which consists of a governing singular quantifier $(a\mathbf{x})$ and a matrix P where $\mathbf{x}$ is any variable and a is any proper name, provided that (i) a does not occur in any undischarged premiss, and (ii) P does not contain a or any dummy name, we may write as a new line the formula which is the same as l except that the singular quantifier $(a\mathbf{x})$ is replaced by the universal quantifier $(\forall \mathbf{x})$. Symbolically:

(UG*) $(a\mathbf{x})P \rightarrow (\forall \mathbf{x})P$ provided that (i) a does not occur in any premiss, and (ii) P does not contain a or any dummy name.

Our explanation of the rule UG* will be postponed until the two remaining rules have been stated and some examples of quantificational deductions have been given.

5. The rules SD* and SQ*; the restriction in SQ*. The rules SQ* (singular quantification) and SD* (singular dequantification) are concerned entirely with singular statements. SQ* is concerned with the move from an unquantified to a quantified and SD* with that from a quantified to an unquantified singular statement.

SQ* is the rule that if we have as a line of a deduction a formula P which contains one or more occurrences of a singular term α we may write as a new line the formula which consists of the singular quantifier $(\alpha\mathbf{x})$ followed by a formula $P^{\alpha:\mathbf{x}}$ which is the same as P except that $\mathbf{x}$ is substituted for α at any number of occurrences; the variable $\mathbf{x}$ that is used must be one that is not captured by any quantifier within P. We express this rule symbolically thus:

(SQ*) $P \rightarrow (\alpha\mathbf{x})P^{\alpha:\mathbf{x}}$, provided that **x** is not captured by any quantifier within P.

In virtue of this rule if we have as a line of a deduction the statement *Socrates is a philosopher* we may write as a new line the statement (*Socrates x*) [*x is a philosopher*]. For simple cases such as this, when P contains no quantifier, the validity of the rule is obvious. In order to see the point of the condition that the variable used must not be captured by any quantifier within P let us consider the statement: *Socrates is a philosopher and everyone admires him.* In quantificational form this might be written:

(1) Socrates is a philosopher. $(\forall x)$ [x admires Socrates]. Suppose that we wish to apply SQ* to (1), taking (1) as P and *Socrates* as α. The following statements, (2), (3), (4), among others, could be deduced from (1) by a correct use of SQ:

(2) (Socrates y)[y is a philosopher. $(\forall x)$[x admires y]].

(3) (Socrates y)[Socrates is a philosopher. $(\forall x)$[x admires y]].

(4) (Socrates x)[x is a philosopher. $(\forall x)$[x admires Socrates]].

(5) and (6) following, however, cannot be obtained from (1) by correct uses of SQ*.

(5) (Socrates x)[Socrates is a philosopher. $(\forall x)$[x admires x]].

(6) (Socrates x)[x is a philosopher. $(\forall x)$[x admires x]]. In both these cases *Socrates* has been replaced at its second occurrence by a variable, x, which is captured by a quantifier, $(\forall x)$, within (1) and the condition of the rule has thus been infringed. The meaning of (5) and (6) is 'Socrates is

such that . . . and everyone admires himself' whereas what we aim to get by SQ*, if the second occurrence of *Socrates* is replaced by a variable, is something which means: 'Socrates is such that . . . and everyone admires *him*'. This meaning is given by either (2) or (3), depending on whether the clause represented by the dots is 'he is a philosopher' or 'Socrates is a philosopher'. The principle is that when a singular term is extracted by SQ* into a quantifier and replaced by a variable that variable must be bound by that quantifier; otherwise the meaning will be altered in a substantial way.

SD* is the rule that if we have a line of deduction which consists of a governing singular quantifier $(\alpha\mathbf{x})$ and a matrix P we may write as a new line the formula which results from omitting the quantifier and replacing $\mathbf{x}$ at every occurrence in P by α. We symbolize this rule as follows:

(SD*) $(\alpha\mathbf{x})P \rightarrow P^{x|\alpha}$.

For example by SD we may go from:

(Socrates x)[x is a philosopher],

to:

Socrates is a philosopher;

and from:

(2) (Socrates y)[y is a philosopher. $(\forall x)$[x admires y]],

to:

Socrates is a philosopher and $(\forall x)$[x admires Socrates]. Notice that we could not in accordance with SD go from (2) to, e.g.,

(7) Socrates is a philosopher and $(\forall x)$[x admires y], since SD requires the variable (y in this case) to be replaced by the singular term (*Socrates* in this case) at every occurrence. (7) is not in any case a proper statement. It

is a pronominal clause with free variable y and can form part of a statement only if preceded by a quantifier which binds the y.

The validity of the rule SD* is obvious if the meanings of the symbols involved are understood.

6. Examples. Now let us have some more extended examples of deductions in which these rules are used. In these examples (1 to 4 in this section) we shall take a to be the name of any individual and $F(\ )$, $G(\ )$, $H(\ )$ to represent any predicates which do not contain within themselves any singular term or any quantifier.

Example 1

Pr × 1

(1) $(\forall x)Fx$

1, UI* × 2[1]

(2) $(ax)Fx$

2, EG* × 3

(3) $(\exists x)Fx$

Example 2

Pr × 2

(1) $(\forall x)[Fx \supset Gx]$

Pr × 2

(2) Fa

1, UI* × 3

(3) $(ax)[Fx \supset Gx]$

3, SD* × 4

(4) $Fa \supset Ga$

4, 2, MP × 5

(5) Ga

5, SQ* × 6

(6) $(ax)Gx$

[1] In each deduction the indented lines are justification lines. See TFL, page 71.

Example 3

Pr × 1

(1) $(\forall x)[Fx \supset Gx]$

Pr × 2

(2) $(\exists x) \sim Gx$

2, EI* × 3

(3) $(dx) \sim Gx$

3, SD* × 4

(4) $\sim Gd$

1, UI* × 5

(5) $(dx)[Fx \supset Gx]$

5, SD* × 6

(6) $Fd \supset Gd$

6, 4, MT × 7

(7) $\sim Fd$

7, SQ* × 8

(8) $(dx) \sim Fx$

8, EG* × 9

(9) $(\exists x) \sim Fx$

Example 4

Pr × 1

(1) $(\forall x)[Fx \supset Gx]$

Pr × 2

(2) $(\forall x)Fx$

1, UI* × 3

(3) $(ax)[Fx \supset Gx]$

3, SD* × 4

(4) $Fa \supset Ga$

2, UI* × 5

(5) $(ax)Fx$

5, SD* × 6

(6) Fa

4, 6, MP × 7

(7) Ga

7, SQ* × 8

(8) $(ax)Gx$

8, UG* × 9

(9) $(\forall x)Gx$

7. Explanatory remarks about UG*[1]. So far we have said nothing by way of justification of the rule UG*. A complete proof of the validity of UG* cannot conveniently be given except in a rather technical form. However, the principle underlying the rule can be stated fairly easily and can be illustrated in a simple case. The principle broadly is that what can be inferred about a certain individual from premisses which contain no special information about that

[1] Although the sets of rules considered differ in some respects from those used in this book the following articles in *Analysis* may be found helpful in the present context on the question of principle: Vol. 19 (1958), J. L. Mackie 'The rules of natural deduction'; Vol. 22 (1962), Robert Price 'Arbitrary symbols and natural deduction'.

individual may be inferred about any other individual also; more strictly it is that what can be inferred about an individual, in a statement in which that individual is denoted by a certain proper name a, from premisses which do not contain that proper name, can be inferred about any other individual also. The conditions of this principle are illustrated in *example* 4. Here we have deduced a statement, (8) $(ax)Gx$, about the individual a, under the name a, from premisses which do not contain that name. Let us suppose that we are here concerned with a universe of discourse which consists of a and a number of other distinct individuals, $a_1, a_2, \ldots, a_i, \ldots$. We can see that lines (3) to (8) of the example 4 deduction can be repeated, with suitable modifications, over and over again, to give us in turn lines (8_1) $(a_1x)Gx$, (8_2) $(a_2x)Gx$, ..., (8_i) $(a_ix)Gx$, ... The deduction is set out schematically below[1] with the lines mentioned occurring at the bottoms of distinct columns but it is to be understood that what is represented is a single deduction in which line (3_1) follows line (8), line (3_2) follows line (8_1) and so on.

The rule UG* allows us in this case to go directly from (8) to (9) $(\forall x)Gx$ and in view of the possible extension of the deduction after line (8) as shown above we can see why: if the premisses (1) and (2) are true then not only (8) but each of the lines (8_1), (8_2), ..., (8_i), ..., must be true; so Gx is true of each of the individuals $a, a_1, a_2, \ldots, a_i, \ldots$ as x; i.e. Gx is true of every individual in the universe as x; in other words $(\forall x)Gx$ is a true statement.

UG* allows us when we have a line $(ax)P$ to write as a new line $(\forall x)P$ provided that three conditions are satisfied: (i) a does not occur in any undischarged premiss; (ii) P does not contain a, and (iii) P does not contain a dummy name. We shall now try to indicate why these conditions are included in the rule. (i) is included to ensure that no special information about a, under that name, is given in

[1] page 47.

$\text{Pr} \times 1$				
(1) $(\forall x)[Fx \supset Gx]$				
$\text{Pr} \times 2$				
(2) $(\forall x)Fx$				
$1, \text{UI}^* \times 3$	$1, \text{UI}^* \times 3_1$		$1, \text{UI}^* \times 3_i$	
(3) $(ax)[Fx \supset Gx]$	(3_1) $(a_1x)[Fx \supset Gx]$	...	(3_i) $(a_ix)[Fx \supset Gx]$	...
$3, \text{SD}^* \times 4$	$3_1, \text{SD}^* \times 4_1$		$3_i, \text{SD}^* \times 4_i$	
(4) $Fa \supset Ga$	(4_1) $Fa_1 \supset Ga_1$	...	(4_i) $Fa_i \supset Ga_i$	...
$2, \text{UI}^* \times 5$	$2, \text{UI}^* \times 5_1$		$2, \text{UI}^* \times 5_i$	
(5) $(ax)Fx$	(5_1) $(a_1x)Fx$	...	(5_i) $(a_ix)Fx$	...
$5, \text{SD}^* \times 6$	$5_1, \text{SD}^* \times 6_1$		$5_i, \text{SD}^* \times 6_i$	
(6) Fa	(6_1) Fa_1	...	(6_i) Fa_i	...
$4, 6, \text{MP} \times 7$	$4_1, 6_1, \text{MP} \times 7_1$		$4_i, 6_i, \text{MP} \times 7_i$	
(7) Ga	(7_1) Ga_1	...	(7_i) Ga_i	...
$7, \text{SQ}^* \times 8$	$7_1, \text{SQ}^* \times 8_1$		$7_i, \text{SQ}^* \times 8_i$	
(8) $(ax)Gx$	(8_1) $(a_1x)Gx$	...	(8_i) $(a_ix)Gx$	...

the premisses. It is useful to compare example 4 with example 2. In example 2 we obtain a line (6) $(ax)Gx$. But, if we try, we will find that we cannot go on therefrom to obtain lines (6_1) $(a_1x)Gx$, (6_2) $(a_2x)Gx$, ... We could indeed obtain these lines if we were able to obtain, corresponding to line (2) Fa, which is used in the deduction of (5), lines (2_1) Fa_1, (2_2) Fa_2, and so on. But since line (2) is a premiss we could obtain these lines only by introducing additional premisses, in which case (6_1) ... would not be obtained from just premisses (1), and (2). In fact in example 2 we have obtained $(ax)Gx$ from premisses which do contain the name a, i.e. give special information about a, and for this reason are unable to obtain lines (6_1), (6_2), ... from these premisses. We can thus see why in this case we are not permitted by the rule UG* to generalize on (6) and so obtain (7) $(\forall x)Gx$.

The reason why condition (ii) (that P does not contain a) is included in UG* is not so immediately evident. Let us suppose, however, that in a deduction we obtain a line (l) Haa where a does not occur in any premiss. We might proceed to obtain by SQ* a subsequent line (m) $(ax)Hax$. This deduction would be perfectly legitimate since SQ* does not require the substitution of x for a at every occurrence within Haa. Because of condition (ii) the rule UG* would not allow us to go on to ($m+1$) $(\forall x)Hax$. We can see the justification if we consider that we would not in general be able to extend our deduction beyond line (m) to obtain (m_1) $(a_1x)Hax$, (m_2) $(a_2x)Hax$, ... and so on in the same sort of way as in example 4 where we were able to obtain lines (8_1) $(a_1x)Gx$, (8_2) $(a_2x)Gx$, ... and so on; and we would not in general be able to obtain these lines (m_1), (m_2), ... because we would not in general be able to obtain the lines (l_1) Haa_1, (l_2) Haa_2, ... from which they would be deduced. We could indeed in general obtain (l_1) Ha_1a_1, (l_2) Ha_2a_2. ... , and hence (m_1) $(a_1x)Ha_1x$, (m_2) $(a_2x)Ha_2x$, ... but the fact that each of these latter could be obtained is no justification for $(\forall x)Hax$.

Why is condition (iii) included? Let us suppose that in a deduction we obtain a line (m) $(ax)Hdx$, where d is a dummy name and a does not occur in the premisses. If we could obtain in addition lines (m_1) $(a_1x)Hdx$, (m_2) $(a_2x)Hdx$ and so on we would be justified in asserting $(\forall x)Hdx$. However, we cannot in general obtain in addition the lines in question. The dummy name d will normally have been introduced by EI*. Let us suppose that the use of EI* which introduced d yielded a line (k) $(dx)[—<a>—]$ which contains d and may or may not contain a. In an extended deduction aiming at (m_1) $(a_1x)Hdx$, (m_2) $(a_2x)Hdx$, and so on we should require, corresponding to (k), (k_1) $(dx)[—<a_1>—]$, (k_2) $(dx)[—<a_2>—]$, and so on, all obtained by means of EI*. However, the use of EI* to obtain any of these lines other than (k) would be illegitimate; for the dummy name brought in by EI* must be one not previously used; hence, since (k_1), (k_2), ... are all later than (k), d cannot be correctly introduced at any of these lines by EI*. We could of course obtain (k_1) $(d_1x)[—<a_1>—]$, (k_2) $(d_2x)[—<a_2>—]$ ... and thence go on to obtain (m_1) $(a_1x)Hd_1x$, (m_2) $(a_2x)Hd_2x$, and so on but our ability to obtain these lines is of course no justification for the assertion of $(\forall x)Hdx$.

These explanatory remarks about UG* may be summed up in a rough-and-ready way thus. Universal generalization from a statement about a single individual is possible because in some cases what can be proved of one individual can be proved of every other individual also. Conditions (i), (ii) and (iii) are inserted to prevent universal generalization in cases where what can be proved of one individual cannot be proved of every other individual also.

We add an example which gives another illustration of the use of UG*. A single pronoun variable (e.g. x) could of course be used throughout this example without any alteration to meaning.

Example 5

	Pr × 1
(1)	$(\forall x)[Fx \supset Gx]$
	Pr × 2
(2)	$(\forall y)[Gy \supset Hy]$
	1, UI* × 3
(3)	$(ax)[Fx \supset Gx]$
	2, UI* × 4
(4)	$(ay)[Gy \supset Hy]$
	3, SD* × 5
(5)	$Fa \supset Ga$
	4, SD* × 6
(6)	$Ga \supset Ha$
	5, 6, HS × 7
(7)	$Fa \supset Ha$
	7, SQ* × 8
(8)	$(az)[Fz \supset Hz)$
	8, UG* × 9
(9)	$(\forall z)[Fz \supset Hz]$

In this example the symbols a, $F(\,)$, $G(\,)$ and $H(\,)$ may be regarded as subject to the condition stated in the second sentence of section 6.

8. Replacement of starred rules by unstarred rules: EI, EG, UI, UG. The quantified singular statement of the form $(a\mathbf{x})F\mathbf{x}$ has been introduced in this book in the belief, right or wrong, that it simplifies the understanding of certain aspects of quantification theory. However, quantified singular statements are not a recognized element in standard quantificational logic and, if we confine our attention to arguments of which no quantified singular statement is a premiss or conclusion, it is possible, and indeed desirable, to replace the six starred rules: EI*, EG*, UI*, UG*, SQ*, SD*, by four unstarred rules: EI, EG, UI, UG, which can be derived from them. The unstarred rules will be stated in abbreviated form; $P^{A|B}$ is used to mean the formula which is the same as P except that A is replaced by B at every free occurrence[1] and $P^{A:B}$ is used to mean a formula which is the same as P except that A is replaced by B at any number of free occurrences. a must be a proper name and α either a proper or a dummy name. The rules are:

(EI) $(\exists \mathbf{x})P \rightarrow P^{\mathbf{x}|\mathbf{d}}$, where **d** is a dummy name not previously used.

(EG) $P \rightarrow (\exists \mathbf{x})P^{\alpha:\mathbf{x}}$, where **x** is a variable not captured by any quantifier within P.

(UI) $(\forall \mathbf{x})P \rightarrow P^{\mathbf{x}|\alpha}$.

(UG) $P \rightarrow (\forall \mathbf{x})P^{a|\mathbf{x}}$, where:

(i) a does not occur in any undischarged premiss;

(ii) P does not contain any dummy name, and

(iii) **x** is not captured by any quantifier within P.

It can easily be shown that in a deduction

[1] Instead of this notation we could have used the following, which is in fact employed in some other monographs in this series: $\dot{S}^{y_1, y_2 \ldots}_{x_1 x_2 \ldots}(P)$ for *the formula which results from substituting* y_1 *for* x_1, y_2 *for* x_2, ... *at all free occurrences in P*. The $P^{x|y}$ notation, though less satisfactory where several substitutions have to be made at once, has some affinity with that used in TFL. see TFL, pages 74–5.

(i) any line correctly deduced by EI can be deduced without EI in two steps, by EI* and SD*;

(ii) any line correctly deduced by EG can be deduced without EG in two steps, by SQ* and EG*;

(iii) any line correctly deduced by UI can be deduced without UI in two steps, by UI* and SD*; and

(iv) any line correctly deduced by UG can be deduced without UG in two steps, by SQ* and UG*.

We give proofs for (i) and (iv). Proofs of (ii) and (iii) are left as exercises to the reader.

(i) In any use of EI a line $P^{\mathbf{x}|\mathbf{d}}$ is deduced from a line $(\exists\mathbf{x})P$, **d** being a dummy name which occurs for the first time in the line $P^{\mathbf{x}|\mathbf{d}}$. Consider a deduction in which $(\exists\mathbf{x})P$ is line (l) and $P^{\mathbf{x}|\mathbf{d}}$ is derived by EI at line (m). This move can be made without EI in two steps after line (m — 1) by EI* and SD*, as follows:

. . .

. . .

(l) $(\exists\mathbf{x})P$

. . .

. . .

l, EI* × m'

(m') $(\mathbf{dx})P$

m', SD* × m

(m) $P^{\mathbf{x}|\mathbf{d}}$.

It may be helpful to have an actual example: let Oy mean *y is an officer* and let By mean *y was born in Britain.* The rule EI allows us to go from $(\exists y)\,[Oy \,.\, By]$ to $Od_1 \,.\, Bd_1$, provided that d_1 has not previously been used. But without EI this move could be made by the starred rules EI* and SD*. For this example the deduction is:

. . .

. . .

(l) $(\exists y)\,[Oy \,.\, By]$

. . .

. . .

l, EI* $\times\, m'$

(m') $(d_1 y)\,[Oy \,.\, By]$

m', SD* $\times\, m$

(m) $Od_1 \,.\, Bd_1$.

(iv) In any use of UG a line $(\forall \mathbf{x})P^{a|\mathbf{x}}$ is derived from a line P, P, a, $\mathbf{x}$ being such that (a) a does not occur in any undischarged premiss, (b) P does not contain any dummy name, (c) $\mathbf{x}$ is a variable which is not captured by any quantifier in P when it is substituted for a at every occurrence in P. Consider a deduction in which P is line (l) and $(\forall \mathbf{x})P^{a|\mathbf{x}}$ is derived at line (m). This move can be made without UG in two steps by SQ* and UG* as follows:

. . .

. . .

(l) P

. . .

. . .

l, SQ* $\times\, m'$

(m') $(a\mathbf{x})P^{a|\mathbf{x}}$

m', UG* $\times\, m$

(m) $(\forall \mathbf{x})P^{a|\mathbf{x}}$.

In this deduction (m') is correctly obtained by SQ*. SQ* allows us to obtain $P^{a:\mathbf{x}}$ (P with $\mathbf{x}$ for a at any number of occurrences), but $P^{a|\mathbf{x}}$ (P with $\mathbf{x}$ for a at every occurrence) is just a particular case of $P^{a:\mathbf{x}}$; further, in virtue of (c) $\mathbf{x}$ is not captured when substituted for a

at every occurrence in P. Line (m) is correctly obtained by UG*. UG* may be applied to (m') provided that (1) a does not occur in any undischarged premiss; (2) $P^{a|\mathbf{x}}$ does not contain a; (3) $P^{a|\mathbf{x}}$ does not contain any dummy name. Conditions (1) and (3) are satisfied in virtue of (a) and (b) above respectively. But $P^{a|\mathbf{x}}$ cannot contain a since it is the formula which is the same as P except that $\mathbf{x}$ is substituted for a at every occurrence; hence condition (2) is satisfied also.

In view of (i), (ii), (iii) and (iv) it is evident that the unstarred rules EI, EG, UI and UG must be valid on the assumption that the six starred rules are valid.[1]

We now illustrate further the use of the unstarred rules by giving in examples (1′) to (5′) below deductions which correspond to those in examples (1) to (5) in sections 6 and 7 except that the unstarred instead of the starred rules are used.

Example 1′

Pr × 1
(1) $(\forall x)Fx$
1, UI × 2
(2) Fa
2, EG × 3
(3) $(\exists x)Fx$

[1] On the question of the validity of our system of rules as a whole it should be said that in this book, as in TFL, the principle of bivalence is adopted as a fundamental postulate; see TFL, page 2.

by unstarred rules: EI, EG, UI, UG

Example 2′

Pr × 1
(1) $(\forall x)[Fx \supset Gx]$

Pr × 2
(2) Fa
1, UI × 3
(3) $Fa \supset Ga$
3, 2,MP × 4
(4) Ga

To obtain the conclusion $(ax)Gx$ of example 2 we should require a starred rule.

Example 3′

Pr × 1
(1) $(\forall x)[Fx \supset Gx]$

Pr × 2
(2) $(\exists x) \sim Gx$
2, EI × 3
(3) $\sim Gd$
1, UI × 4
(4) $Fd \supset Gd$
4, 3, MT × 5
(5) $\sim Fd$
5, EG × 6
(6) $(\exists x) \sim Fx$

Example 4′

Pr × 1
(1) $(\forall x)[Fx \supset Gx]$

Pr × 2
(2) $(\forall x)Fx$
1, UI × 3
(3) $Fa \supset Ga$
2, UI × 4
(4) Fa
3, 4, MP × 5
(5) Ga
5, UG × 6
(6) $(\forall x)Gx$

Example 5′

Pr × 1
(1) $(\forall x)[Fx \supset Gx]$

Pr × 2
(2) $(\forall y)[Gy \supset Hy]$
1, UI × 3
(3) $Fa \supset Ga$
2, UI × 4
(4) $Ga \supset Ha$
3, 4, HS × 5
(5) $Fa \supset Ha$
5, UG × 6
(6) $(\forall z)[Fz \supset Hz]$

CHAPTER THREE

1. List of rules. In this section we set out for reference all the deductive rules, other than those concerned with identity and the iota operator,[1] which we adopt and may make use of in the remainder of the book. Some of these rules could, of course, be derived from others; in particular the eight quantificational equivalences XII to XIX listed under (10) can be proved on the basis of the remainder of the rules and will not be used until after proofs of some of them have been given on pages 71–73. The rule TF will not be used until after an explanatory remark has been made on page 68.

The rules are to be interpreted in the light of remarks which will be made in section 2.

(1) Premiss rule (Pr). Any formula may be written as a line of a deduction and marked as a premiss with the sign ┌.

(2) *Modus ponens* (MP). $P \supset Q, P \rightarrow Q.$

(3) *Modus tollens* (MT). $P \supset Q, \sim Q \rightarrow \sim P.$

(4) Hypothetical syllogism (HS).
$P \supset Q, Q \supset R \rightarrow P \supset R.$

(5) *Modus tollendo ponens* (MTP). $P \vee Q, \sim P \rightarrow Q.$

(6) Addition (Add). $P \rightarrow P \vee Q.$

(7) Simplification (Simp). $P.Q \rightarrow P; P.Q \rightarrow Q.$

[1] Identity and the iota operator are discussed in sections 7 and 8, respectively, of the present chapter. The rules for identity are given on page 105 and those for the iota operator on page 110.

(8) Conjunction (Conj). $P, Q \rightarrow P.Q$.

(9) Rule of conditional proof (CP).
$P_1, P_2, \ldots, P_{n-1}, (P_1, P_2, \ldots, P_{n-1}, P_n \vdash Q) \rightarrow P_n \supset Q$.

(10) Rule of interchange (Int).
$P \rightarrow P^{K:K'}$, where K is a truth-functional component of P and either $K \equiv K'$ or $K' \equiv K$ is an exemplification of one of the logical equivalences in the following list.

I. De Morgan's Laws
$$\sim[P.Q] \equiv \sim P \vee \sim Q$$
$$\sim[P \vee Q] \equiv \sim P . \sim Q$$

II. Laws of Commutation
$$P.Q \equiv Q.P$$
$$P \vee Q \equiv Q \vee P$$

III. Laws of Association
$$P.[Q.R] \equiv [P.Q].R$$
$$P \vee [Q \vee R] \equiv [P \vee Q] \vee R$$

IV. Laws of Distribution
$$P.[Q \vee R] \equiv [P.Q] \vee [P.R]$$
$$P \vee [Q.R] \equiv [P \vee Q].[P \vee R]$$

V. Law of Double Negation
$$P \equiv \sim\sim P$$

VI. Law of Exportation and Importation
$$[P.Q] \supset R \equiv P \supset [Q \supset R]$$
$$P \supset [Q \supset R] \equiv Q \supset [P \supset R]$$

VII. Law of transposition
$$P \supset Q \equiv \sim Q \supset \sim P$$
$$P \supset \sim Q \equiv Q \supset \sim P$$
$$\sim P \supset Q \equiv \sim Q \supset P$$

VIII.
$$P \vee Q \equiv \sim P \supset Q$$
$$P \supset Q \equiv \sim P \vee Q$$
$$P \supset Q \equiv \sim[P . \sim Q]$$

IX. $P \equiv Q \equiv [P \supset Q].[Q \supset P]$

$P \equiv Q \equiv [P.Q] \vee [\sim P. \sim Q]$

X. $P \vee P \equiv P$

$\sim P \supset P \equiv P$

$P \supset \sim P \equiv \sim P$

XI. $P.P \equiv P$

XII. $(\forall \mathbf{x})F\mathbf{x} \equiv \sim(\exists \mathbf{x}) \sim F\mathbf{x}$

XIII. $(\exists \mathbf{x})F\mathbf{x} \equiv \sim(\forall \mathbf{x}) \sim F\mathbf{x}$

XIV. $(\forall \mathbf{x}) \sim F\mathbf{x} \equiv \sim(\exists \mathbf{x})F\mathbf{x}$

XV. $(\exists \mathbf{x}) \sim F\mathbf{x} \equiv \sim(\forall \mathbf{x})F\mathbf{x}$

XVI. $(\forall \mathbf{x})[F\mathbf{x} \supset G\mathbf{x}] \equiv \sim(\exists \mathbf{x})[F\mathbf{x}. \sim G\mathbf{x}]$

XVII. $(\exists \mathbf{x})[F\mathbf{x}.G\mathbf{x}] \equiv \sim(\forall \mathbf{x})[F\mathbf{x} \supset \sim G\mathbf{x}]$

XVIII $(\forall \mathbf{x})[F\mathbf{x} \supset \sim G\mathbf{x}] \equiv \sim(\exists \mathbf{x})[F\mathbf{x}.G\mathbf{x}]$

XIX. $(\exists \mathbf{x})[F\mathbf{x}. \sim G\mathbf{x}] \equiv \sim(\forall \mathbf{x})[F\mathbf{x} \supset G\mathbf{x}]$

(11) Truth-functional deduction (TF).
$P_1, P_2, \ldots, P_n \rightarrow Q$, where $P_1, P_2, \ldots, P_n$; *therefore* Q is a truth-functionally valid argument.

(12) Existential instantiation (EI).
$(\exists \mathbf{x})P \rightarrow P^{\mathbf{x}|\mathbf{d}}$, where **d** is a dummy name not previously used.

(13) Existential generalization (EG).
$P \rightarrow (\exists \mathbf{x})P^{\alpha:\mathbf{x}}$, where α is a proper name or dummy name and **x** is not captured by any quantifier within P.[1]

(14) Universal instantiation (UI).
$(\forall \mathbf{x})P \rightarrow P^{\mathbf{x}|\alpha}$, where α is a proper name or dummy name.

[1] See also reference to this rule on page 110.

(15) Universal generalization (UG).

$P \rightarrow (\forall \mathbf{x})P^{a|\mathbf{x}}$, where a is a proper name and (i) a does not occur in any undischarged premiss, (ii) P does not contain any dummy name, and (iii) $\mathbf{x}$ is not captured by any quantifier within P.

2. Interpretation of the rules. Many of the rules set forth in section 1 are truth-functional and correspond to rules used in TFL. However, it is intended that the applicability of the present set of rules shall be different from that of the rules in TFL in the following respect. In TFL each rule could be applied to either statements or statement forms; for example the rule of simplification could be applied directly either to a statement, e.g. *Dogs bark . lions roar*, or to a statement form, e.g. $p.q$. The rules which have just been listed (including the rule of simplification) on the other hand are to be applied directly only to statements. Despite this difference we may still allow ourselves to construct deductions in which some or all of the lines are not statements but statement forms. What we call deductions may be divided into two groups: (i) actual deductions, (ii) deduction schemata. In an actual deduction every line is a statement whereas in a deduction schema one or more lines are statement forms; in the examples in section 3 the deductions in examples (i) and (v) are actual deductions and the remainder are deduction schemata; in (x) we should have an actual deduction if in the list of abbreviations a were included as standing for a designated proper name. An actual deduction is of course correctly constructed if each step is taken in accordance with one of the listed rules. A deduction schema on the other hand may be regarded as correctly constructed if every actual deduction which exemplified the schema would be correctly constructed. An actual deduction exemplifies a deduction schema if it is formed from the schema by consistent replacement of the variables, other than pronoun variables, in such a way that at every occurrence the same predicate variable is replaced by the same predicate, the same proper name variable by the same proper name and the same statement variable by the same statement.[1] It should be added that in some cases when a statement form is written down, whether or not as part of a deduction schema, some condition is added restricting the choice of replacements for one or more of the variables. When this occurs in the context of a deduction schema the schema is of course to be regarded as correctly constructed if every exemplification of the schema which satisfied the conditions in question would be correctly constructed. An instance of a condition of this kind is the note in small type immediately below example (iv) which restricts the choice of replacements for the variable a; in any exemplification of this schema the replacement for a must not be contained in the replacement for $F(\ )$ or $G(\ )$.

Two important uses of the deductive method are to prove the validity of valid arguments or argument forms and to prove the logical truth of logically true statements or statement forms. The existence of dummy names gives rise to a slight complication in this connexion. Let us call a deduction (whether an actual deduction or a deduction schema) *complete* if it is correctly constructed and does not contain in an undischarged premiss or in its last line any dummy name. An argument form is proved to be valid and a statement form is proved to be logically true by

[1] It would be possible to distinguish between variables which do and those which do not occur in an undischarged premiss or the last line of a complete deduction schema, and to state a requirement for correct construction which would be less stringent in respect of the latter than of the former. However, it seems unnecessary to introduce this complication here.

means of a complete deduction schema. The validity of an argument or the logical truth of a statement can be proved by means of a complete actual deduction, but in many cases it is natural to use a deduction schema for this purpose also. An instance is example (xxi), which is a deduction schema and not an actual deduction in virtue solely of the occurrence as a proper name variable of the symbol a_1.

3. Principles of single quantification; deductions. We now proceed to give a number of examples illustrating deductive methods. In some cases we construct deductions to prove the validity of actual arguments which have themselves no particular interest of a general kind. However, for the most part a deduction will be related to a logical principle of some general importance. By a logical principle we mean in the present context a quantificationally valid argument form or a logically true quantificational statement form. We shall set forth a number of lists of logical principles and after each list give deductions establishing the validity of some of the principles in the list. Alongside the principles a number of other forms will be set out in a right-hand column; these forms are in some sort of family relationship to the principles but are fallacious. Proofs of their fallaciousness will, in some cases at least, be given in chapter four, section 6.

From what has been said in the previous section it should be evident that checking for correct construction is a less straightforward matter in the case of a deduction schema than of an actual deduction. But it is of course of fundamental importance to be able to follow the direct application of the rules to statements in actual deductions, and the beginner who experiences difficulty might be well advised to start by transforming the deduction schemata in our examples into actual deductions and checking the application of the rules in these. To transform a deduction schema into an actual deduction we have to make suitable substitutions for the non-pronoun variables consistently throughout the schema. For instance, in example (ii) we could get an actual example by substituting for *F*( ) (   ) *was a ticket-holder*, for *G*( ) (   ) *was admitted*, and for *a Smith*.[1] In the first instance the student should replace predicate variables by predicates which do not contain within themselves either any proper name or any quantifier; if the resulting deduction proves to be correctly constructed he might go on to consider, in the case of any deduction which made use of the rule UG, whether it would still be correct if these predicates were replaced by others of which at least one contained a proper name, and in the case of a deduction in which UG or EG was used to introduce an **x**-quantifier whether it would still be correct if these predicates were replaced by others of which at least one contained within itself an **x**-quantifier. After some practice of this kind he should be able to go on to check the correctness of deduction schemata themselves.

Logical principles	Fallacies
(1) $(\forall x)[Fx \supset Gx]$, Fa; $\therefore$ Ga	(1.1) $(\forall x)[Fx \supset Gx]$, Ga; $\therefore$ Fa
(2) $(\forall x)[Fx \supset Gx]$, $\sim Ga$; $\therefore$ $\sim Fa$	(2.1) $(\forall x)[Fx \supset Gx]$, $\sim Fa$; $\therefore$ $\sim Ga$
(3) $(\forall x)[Fx \equiv Gx]$, Fa; $\therefore$ Ga	
(4) $(\forall x)[Fx \equiv Gx]$, $\sim Fa$; $\therefore$ $\sim Ga$	
(5) $(\forall x)[Fx \equiv Gx]$, $\sim Ga$; $\therefore$ $\sim Fa$	
(6) $(\forall x)[Fx \equiv Gx]$, Ga; $\therefore$ Fa	

Example i

The following argument is valid:

Anyone is eligible for this appointment if, but only if,

[1] For predicate substitution see chapter 1, section 8.

he is a ratepayer or the son of a ratepayer. Smith is a ratepayer. Therefore he is eligible.

With abbreviations:

$E(\ )$ for ( ) *is eligible for this appointment*, $R(\ )$ for ( ) *is a ratepayer*, $S(\ )$ for ( ) *is the son of a ratepayer*, a for *Smith*,

the argument may be expressed thus:

$(\forall x)[[Rx \vee Sx] \equiv Ex]$, Ra; therefore Ea.

It is shown to be valid by means of the following deductions:

Pr × 1
(1) $(\forall x)[[Rx \vee Sx] \equiv Ex]$

Pr × 2
(2) Ra

1, UI × 3
(3) $[Ra \vee Sa] \equiv Ea$

3, int (IX) × 4
(4) $[[Ra \vee Sa] \supset Ea] . [Ea \supset [Ra \vee Sa]]$

4, simp × 5
(5) $[Ra \vee Sa] \supset Ea$

2, add × 6
(6) $Ra \vee Sa$

5, 6, MP × 7
(7) Ea

Example ii

(2) $(\forall x)[Fx \supset Gx]$, $\sim Ga$; *therefore* $\sim Fa$ is shown to be valid by the following deduction:

Pr × 1
(1) $(\forall x)[Fx \supset Gx]$

Pr × 2
(2) $\sim Ga$
1, UI × 3
(3) $Fa \supset Ga$
3, 2, MT × 4
(4) $\sim Fa$

Logical principles	Fallacies
(7) $(\forall x)[Fx \supset Gx]$, $(\forall x)Fx$; $\therefore$ $(\forall x)Gx$ (8) $(\forall x)[Fx \supset Gx]$, $(\exists x)Fx$; $\therefore (\exists x)Gx$ (9) $(\forall x)[Fx \supset Gx]$, $\sim(\forall x)Gx$; $\therefore$ $\sim(\forall x)Fx$ (10) $(\forall x)[Fx \supset Gx]$, $\sim(\exists x)Gx$ $\therefore$ $\sim(\exists x)Fx$	(7.1) $(\forall x)[Fx \supset Gx]$, $(\forall x)Gx$; $\therefore$ $(\forall x)Fx$

Example iii

(8) $(\forall x)[Fx \supset Gx]$, $(\exists x)Fx$; *therefore* $(\exists x)Gx$ is shown by the following deduction to be a valid argument form.

Pr × 1
(1) $(\forall x)[Fx \supset Gx]$

Pr × 2
(2) $(\exists x)Fx$
2 EI × 3
(3) Fd
1 UI × 4
(4) $Fd \supset Gd$
4, 3 MP × 5
(5) Gd
5 EG × 6
(6) $(\exists x)Gx$

It should be noted that here if after line (2) we had applied UI to (1) to obtain the present (4) we should not have been able to obtain the present line (3) since EI does not allow the use of a dummy name which has previously occurred. In general, if we have a line governed by a particular quantifier and one or more lines governed by universal quantifiers and we wish to obtain from all of them formulae which are singular in respect of the same dummy name **d**, then we must apply EI to the particular formula before applying UI to any of the universal formulae.

Logical principles	Fallacies
(11) $(\forall x)[Fx \supset Gx]$, $(\forall x)[Gx \supset Hx]$; $\therefore$ $(\forall x)[Fx \supset Hx]$	(11.1) $(\forall x)[Fx \supset Gx]$, $(\forall x)[Hx \supset Gx]$; $\therefore$ $(\forall x)[Fx \supset Hx]$
(12) $(\exists x)[Fx . Gx]$, $(\forall x)[Gx \supset Hx]$; $\therefore$ $(\exists x)[Fx . Hx]$	(12.1) $(\forall x)[Fx \supset Gx]$, $(\exists x)[Gx . Hx]$; $\therefore$ $(\exists x)[Fx . Hx]$
(13) $(\forall x)[Gx \supset Fx]$, $(\forall x)[Gx \supset Hx]$, $(\exists x)Gx$; $\therefore$ $(\exists x)[Fx . Hx]$	(13.1) $(\forall x)[Gx \supset Fx]$, $(\forall x)[Gx \supset Hx]$; $\therefore$ $(\exists x)[Fx . Hx]$
(14) $(\forall x)[Fx \supset Gx]$, $(\forall x)[Gx \supset \sim Hx]$; $\therefore$ $(\forall x)[Fx \supset \sim Hx]$	(14.1) $(\forall x)[Fx \supset \sim Gx]$, $(\forall x)[Gx \supset Hx]$; $\therefore$ $(\forall x)[Fx \supset \sim Hx]$

The argument forms in this list correspond to certain traditional syllogistic forms and may themselves be described as syllogistic. It may be remarked that although (13.1) is invalid the form corresponding to it in the traditional logic (*All g are f*, *All g are h; therefore some f are h*) was regarded as valid. The reason for discrepancies of this kind has been discussed in chapter one, section 10.

3.3. *Principles of single quantification; deductions*

Example iv

(11) $(\forall x)[Fx \supset Gx]$, $(\forall x)[Gx \supset Hx]$; *therefore* $(\forall x)[Fx \supset Hx]$ is shown to be valid by the following deduction.

	Pr × 1		2 UI × 4
(1)	$(\forall x)[Fx \supset Gx]$	(4)	$Ga \supset Ha$
	Pr × 2		3, 4 HS × 5
(2)	$(\forall x)[Gx \supset Hx]$	(5)	$Fa \supset Ha$
	1 UI × 3		5 UG × 6
(3)	$Fa \supset Ga$	(6)	$(\forall x)[Fx \supset Hx]$

a must be a proper name which does not occur in *F*( ), *G*( ) or *H*( )[1]

Example v

All foundation members who received the report protested. Some recipient of the report did not protest. Therefore some recipient of the report is not a foundation member.

Using the abbreviations:

F( ) *for* (   ) is a foundation member, *R*( ) *for* (   ) received the report, *P*( ) *for* (   ) protested,

this argument may be expressed as follows:

$(\forall y)[[Fy . Ry] \supset Py]$, $(\exists x)[Rx . \sim Px]$; therefore $(\exists z)[Rz . \sim Fz]$.

[1] See reference to example iv on page 60.

It is shown to be valid by the following deduction.

Pr × 1
(1) $(\forall y)[[Fy.Ry]\supset Py]$

Pr × 2
(2) $(\exists x)[Rx.\sim Px]$
2 EI × 3
(3) $Rd.\sim Pd$
3 simp × 4, 5
(4) Rd

(5) $\sim Pd$
1 UI × 6
(6) $[Fd.Rd]\supset Pd$
6, 7 MT × 7
(7) $\sim[Fd.Rd]$
7 int (V) × 8
(8) $\sim[Fd.\sim\sim Rd]$
8 int (VIII) × 9
(9) $Fd\supset\sim Rd$
5 int (V) × 10
(10) $\sim\sim Rd$
9, 10 MT × 11
(11) $\sim Fd$
4, 11 conj × 12
(12) $Rd.\sim Fd$
13 EG × 13
(13) $(\exists z)[Rz.\sim Fz]$

In the formal expression of this argument different variables, x, y, z, have been used in the two premisses and the conclusion. This, though legitimate, is unnecessary and the argument would more normally be expressed with a single variable, say x, throughout. Thus the following argument is identical in meaning with the one just shown to be valid:

$(\forall x)[[Fx.Rx]\supset Px]$, $(\exists x)[Rx.\sim Px]$; therefore $(\exists x)[Rx.\sim Fx]$.

The reader may like to check that this argument can be shown to be valid by a deduction which is identical with that just given except that y and z are each replaced at every occurrence by x.

Example (v) may be used as the occasion for a modification to our truth-functional rules. In this example we have used a deduction of thirteen lines. It may be felt that this is longer than should be necessary for a very simple argument. Alternative deductions are indeed possible within our existing rules but it is probable that none of them would be appreciably shorter; so that if we want to have shorter deductions in this and similar cases we must make some modifications to our system of rules. The length of the deduction in example (v) is mainly in the truth-functional part; and an improvement, in respect of length of deductions, could be brought about by various piece-meal additions to our truth-functional rules. For example, we might introduce an alternative form of the rule *modus tollens*: viz. $P \supset \sim Q,\ Q \rightarrow \sim P$. Again, we might add to our list of logical equivalences under VIII the equivalence: $p \supset \sim q \equiv \sim[p.q]$. With these two additions to our rules it can be seen that lines (8) and (10) could be eliminated from the example (v) deduction. However, since we are concerned in this book primarily with the purely quantificational aspect of quantificational arguments, it is preferable to make a change of a more wholesale kind in respect of our purely truth-functional rules. Accordingly we adopt the following rule (for which we use the name *TF* applied by W. V. Quine[1] to a rule that is substantially the same):

(TF) $P_1, P_2, \ldots, P_n \rightarrow Q$, where $P_1, P_2, \ldots, P_n$; *therefore* Q is a truth-functionally valid argument.

This rule, which is shown as (11) in the list on page 58, allows the maximum possible reduction to be made to the truth-functional part or parts of a deduction. It should be observed that although easily stated it is in fact a rule of a much more complex kind than any we have so far had, in that to apply it we have to assure ourselves that a certain argument or argument form is truth-functionally

[1] In *Methods of Logic*, a book in which the reader will find many practical hints, some of which are passed on in the present chapter.

valid. This is often easily done but may be difficult. In a sense the whole of truth-functional logic is taken for granted in this rule. A broad characteristic of deductive systems in general is illustrated here: shorter deductions require more numerous or more complex rules. We adopt TF in addition to our existing truth-functional rules rather than in place of them and may still use some of the more elementary rules from time to time.

With TF available the example (v) deduction could be reduced to the following.

Pr × 1
(1) $(\forall y)[[Fy.Ry] \supset Py]$

Pr × 2
(2) $(\exists x)[Rx. \sim Px]$

2 EI × 3
(3) $Rd. \sim Pd$

1 UI × 4
(4) $[Fd.Rd] \supset Pd$

3, 4 TF × 5
(5) $Rd. \sim Fd$

4 EG × 6
(6) $(\exists z)[Rz. \sim Fz]$

Logical principles	Fallacies
(15) $(\forall x)[Fx \supset Gx] \supset (\forall x)[\sim Gx \supset \sim Fx]$ (16) $(\exists x)[Fx.Gx] \supset (\exists x)[Gx.Fx]$ (17) $(\forall x)Fx \supset (\exists x)Fx$	(15.1) $(\forall x)[Fx \supset Gx] \supset (\forall x)[Gx \supset Fx]$

In examples vi and vii (15) and (17) respectively are shown to be logically true.

Example vi

Pr × 1
(1) $(\forall x)[Fx \supset Gx]$
1, UI × 2
(2) $Fa \supset Ga$
2, int (VII) × 3
(3) $\sim Ga \supset \sim Fa$
3, UG × 4
(4) $(\forall x)[\sim Gx \supset \sim Fx]$

(1⊢4) × 5
(5) $(\forall x)[Fx \supset Gx] \supset (\forall x)[\sim Gx \supset \sim Fx]$

a must be a proper name which does not occur in *F*( ) or *G*( ).

Example vii

Pr × 1
(1) $(\forall x)Fx$
1, UI × 2
(2) Fa
2, EG × 3
(3) $(\exists x)Fx$

(1⊢3) × 4
(4) $(\forall x)Fx \supset (\exists x)Fx$

In example (vii) a modified method, which will be adhered to hereafter, has been used for separating the subsidiary deduction (lines 1 to 3) from the main deduction. Instead of a continuous line as in example (vi) we have a pair of corners, the premiss sign at line (1), and a premiss sign on its back after line 3 to mark the point at which the premiss is discharged.

Before going further we give deductions proving two of

the eight quantificational equivalences numbered XII to XIX under the rule of interchange on page 58. It will be seen that the particular pronoun variable x which we use could be replaced throughout by any other pronoun variable, e.g. y.[1]

Proof of XII

Pr × 1
(1) $(\forall x)Fx$

Pr × 2
(2) $(\exists x) \sim Fx$
2, EI × 3
(3) $\sim Fd$
1, UI × 4
(4) Fd
4, add × 5
(5) $Fd \vee \sim(\exists x) \sim Fx$
5, 3, MTP × 6
(6) $\sim(\exists x) \sim Fx$

1, (1, 2⊢6) × 7
(7) $(\exists x) \sim Fx \supset \sim(\exists x) \sim Fx$
7, int (X) × 8
(8) $\sim(\exists x) \sim Fx$

(1⊢8) × 9
(9) $(\forall x)Fx \supset \sim(\exists x) \sim Fx$

Pr × 10
(10) $\sim Fa$
10, EG × 11
(11) $(\exists x) \sim Fx$

(10⊢11) × 12
(12) $\sim Fa \supset (\exists x) \sim Fx$

Pr × 13
(13) $\sim(\exists x) \sim Fx$
12, 13, TF × 14
(14) Fa
14, UG × 15
(15) $(\forall x)Fx$

(13⊢15) × 16
(16) $\sim(\exists x) \sim Fx \supset (\forall x) Fx$
9, 16, conj × 17
(17) $[(\forall x)Fx \supset \sim(\exists x) \sim Fx] . [\sim(\exists x) \sim Fx \supset (\forall x)Fx]$
17, int (IX) × 18
(18) $(\forall x)Fx \equiv \sim(\exists x) \sim Fx$

a must be a proper name which does not occur in $F(\ )$: the application of *UG* to line 14 to obtain line 15 would not be permissible if *a* occurred in the undischarged premiss (13).

[1] See also remark about (38) and (39) on page 79.

This deduction illustrates the usual method of deducing an equivalence. Since any formula $P \equiv Q$ is equivalent to a conjunction of the two implications $P \supset Q$ and $Q \supset P$ it may be shown to be logically true if the two implications are deduced separately and the rules of conjunction and interchange then applied. In cases of this sort the deductions of the two implications will sometimes be given side by side as above where the left-right implication is deduced in lines 1 to 9 and the deduction of the right-left implication is shown in the right-hand column in lines 10 to 16.

The method used to obtain line 6 should also be noted. Once the two mutually contradictory lines (3) and (4) have been deduced we are able to obtain any proposition we want by the use of addition and *modus tollendo ponens*: what we normally do want is, as here, the negation of the last premiss. However, it should be observed that with the rule TF available we could have gone straight from the contradictory lines (3) and (4) to line (6). We will use this type of short cut in future.

Henceforward we shall use freely any of the equivalences listed under the rule of interchange.

It may be remarked here that other and stronger forms of the rule of interchange are possible. For example the restriction that K is a truth-functional component of P is not essential; interchange is in fact possible also where K is a quantificational component of P. However, we do not attempt to show this here. Again interchange might be allowed in respect of any known logical equivalence whether included in our list or not. However, it would be dangerous to go beyond *logical* equivalence: for example, a variant of the rule which allowed interchange in respect of an equivalence $K \equiv K'$, where this occurred as a premiss or as a deduction from undischarged premisses without being a logical equivalence, would be fallacious.

Proof of XV

Pr × 1
(1) $(\exists x) \sim Fx$
1, EI × 2
(2) $\sim Fd$
Pr × 3
(3) $(\forall x)Fx$
3, UI × 4
(4) Fd
2, 4, TF × 5
(5) $\sim(\forall x)Fx$
1, (1, 3⊢5) × 6
(6) $(\forall x)Fx \supset \sim(\forall x)Fx$
6, TF × 7
(7) $\sim(\forall x)Fx$
(1⊢7) × 8
(8) $(\exists x) \sim Fx \supset \sim(\forall x)Fx$

Pr × 9
(9) $\sim Fa$
8, EG × 10
(10) $(\exists x) \sim Fx$
(9⊢10) × 11
(11) $\sim Fa \supset (\exists x) \sim Fx$
Pr × 12
(12) $\sim(\exists x) \sim Fx$
11, 12, TF × 13
(13) Fa
13, UG × 14
(14) $(\forall x)Fx$
(12⊢14) × 15
(15) $\sim(\exists x) \sim Fx \supset (\forall x)Fx$
15, TF × 16
(16) $\sim(\forall x)Fx \supset (\exists x) \sim Fx$
8, 16, TF × 17
(17) $(\exists x) \sim Fx \equiv \sim(\forall x)Fx$

a must be a proper name not contained in *F*( ).

Logical principles	Fallacies
(18) $(\forall x)[Fx.Gx] \equiv [(\forall x)Fx.(\forall x)Gx]$	
(19) $(\exists x)[Fx.Gx] \supset [(\exists x)Fx.(\exists x)Gx]$	(19.1) Converse of 19
(20) $[(\forall x)Fx \vee (\forall x)Gx] \supset (\forall x)[Fx \vee Gx]$	(20.1) Converse of 20
(21) $(\exists x)[Fx \vee Gx] \equiv [(\exists x)Fx \vee (\exists x)Gx]$	
(22) $(\forall x)[Fx \supset Gx] \supset [(\forall x)Fx \supset (\forall x)Gx]$	(22.1) Converse of 22
(23) $(\exists x)[Fx \supset Gx] \equiv [(\forall x)Fx \supset (\exists x)Gx]$	
(24) $(\forall x)[Fx \supset Gx] \supset [(\exists x)Fx \supset (\exists x)Gx]$	
(25) $(\forall x)[Fx \equiv Gx] \supset [(\forall x)Fx \equiv (\forall x)Gx]$	(25.1) Converse of 25
(26) $(\forall x)[Fx \equiv Gx] \supset [(\exists x)Fx \equiv (\exists x)Gx]$	

3.3. Principles of single quantification; deductions

Formulae of the form $(\exists x)[P \supset Q]$ require some comment. Consider $(\exists x)[Fx \supset Gx]$ which appears as the left-hand side of (23). Let us read Fx here as *x is an f* and Gx as *x is a g*. Since $(\forall x)[Fx \supset Gx]$ means *Every f is a g* the beginner is often tempted to interpret $(\exists x)[Fx \supset Gx]$ as meaning *Some f is a g*. However, this is wrong; for *some f is a g* requires that at least one *f* exists, whereas $(\exists x)[Fx \supset Gx]$ can, and indeed must[1], be true if no *f* at all exists. We can see that $(\exists x)[Fx \supset Gx]$ is true if but only if $Fx \supset Gx$ is true of at least one individual *a* as *x*, in other words if it is not the case that $Fx \supset Gx$ is false of every individual *a* as *x*. But $Fx \supset Gx$ is false of *a* as *x* if and only if Fx is true and Gx is false of *a* as *x*. Hence $Fx \supset Gx$ is false of every individual as *x* if and only if Fx is true of every individual and Gx is false of every individual; that is to say, on our interpretation, if and only if everything is an *f* and nothing is a *g*. $(\exists x)[Fx \supset Gx]$ then is true provided that it is not the case that everything is an *f* and nothing is a *g*; otherwise it is false. This last statement gives the essence of what is expressed formally in (23) $(\exists x)[Fx \supset Gx] \equiv [(\forall x)Fx \supset (\exists x)Gx]$, since the right-hand side of (23) is equivalent to $\sim[(\forall x)Fx . \sim(\exists x)Gx]$ and so to $\sim[(\forall x)Fx . (\forall x) \sim Gx]$.

In examples (viii) and (ix) the logical truth of (21) and (22) respectively is proved.

In the left-hand column of the deduction given in example (viii) there is illustrated a common method of establishing a disjunction. We arrive at the disjunction $(\exists x)Fx \vee (\exists x)Gx$ in line (9) by first obtaining the equivalent implication $\sim(\exists x)Fx \supset (\exists x)Gx$.

The deduction of the right-left implication $[(\exists x)Fx \vee (\exists x)Gx] \supset (\exists x)[Fx \vee Gx]$ in the right-hand column is perhaps less obvious than any we have so far had. The antecedent of this implication is naturally written as a premiss at line (11) but there is no obvious direct way

[1] See footnote to page 27.

3.3. Principles of single quantification; deductions

Example viii

Pr × 1

(1) $(\exists x)[Fx \vee Gx]$

1 EI × 2

(2) $Fd \vee Gd$

Pr × 3

(3) $\sim(\exists x)Fx$

3 int (XIV) × 4

(4) $(\forall x) \sim Fx$

4 UI × 5

(5) $\sim Fd$

2, 5 MTP × 6

(6) Gd

6 EG × 7

(7) $(\exists x)Gx$

1 (1, 3 ⊢ 7) × 8

(8) $\sim(\exists x)Fx \supset (\exists x)Gx$

8 int (VIII) × 9

(9) $(\exists x)Fx \vee (\exists x)Gx$

(1 ⊢ 9) × 10

(10) $(\exists x)[Fx \vee Gx] \supset [(\exists x)Fx \vee (\exists x)Gx]$

Pr × 11

(11) $(\exists x)Fx \vee (\exists x)Gx$

11 TF, int × 12

(12) $(\forall x) \sim Fx \supset \sim(\forall x) \sim Gx$

Pr × 13

(13) $\sim(\exists x)[Fx \vee Gx]$

13 int (XIV) × 14

(14) $(\forall x) \sim [Fx \vee Gx]$

14 UI × 15

(15) $\sim [Fa \vee Ga]$

15 TF × 16, 17

(16) $\sim Fa$

(17) $\sim Ga$

16 UG × 18

(18) $(\forall x) \sim Fx$

17 UG × 19

(19) $(\forall x) \sim Gx$

12, 18 MP × 20

(20) $\sim(\forall x) \sim Gx$

19, 20 TF × 21

(21) $(\exists x)[Fx \vee Gx]$

11 (11, 13 ⊢ 21) × 22

(22) $\sim(\exists x)[Fx \vee Gx] \supset (\exists x)[Fx \vee Gx]$

22 TF × 23

(23) $(\exists x)[Fx \vee Gx]$

(11 ⊢ 23) × 24

(24) $[(\exists x)Fx \vee (\exists x)Gx] \supset (\exists x)[Fx \vee Gx]$

10, 24 TF × 25

(25) $(\exists x)[Fx \vee Gx] \equiv [(\exists x)Fx \vee (\exists x)Gx]$

a must be a proper name not contained in *F*( ) or *G*( ).

of obtaining therefrom the consequent. Accordingly we write as an additional premiss (13) the negation, $\sim(\exists x)[Fx \vee Gx]$, of the consequent. By the medium of the two mutually contradictory lines (19) and (20) it is shown that the consequent follows from its own negation. Hence in accordance with the method of indirect deduction it is asserted on its own at line 23

Example ix

Pr × 1
(1) $(\forall x)[Fx \supset Gx]$
1 UI × 2
(2) $Fa \supset Ga$

Pr × 3
(3) $(\forall x)Fx$
3 UI × 4
(4) Fa

2, 4, MP × 5
(5) Ga
5 UG × 6
(6) $(\forall x)Gx$

(3⊢6) × 7
(7) $(\forall x)Fx \supset (\forall x)Gx$

(1⊢7) × 8
(8) $(\forall x)[Fx \supset Gx] \supset [(\forall x)Fx \supset (\forall x)Gx]$

a must be a proper name not contained in *F*( ) or *G*( ).

Here in the justification line for (7) a simplified notation for conditional proof has been adopted. Instead of 1 (1, 3⊢6) × 7 we have written simply (3⊢6) × 7. This might be regarded as an abbreviation for:

> All premisses undischarged before (3) together with the subsidiary deduction *from* all premisses undischarged at (3) *to* (6) yield (7),

which in this case amounts to the same as 1 (1, 3⊢6) × 7. It can be seen that with this interpretation no change is required in our notation for the case where the subsidiary deduction has only one premiss. The justification line for (8), i.e. $(1 \vdash 7) \times 8$, provides an example: since there is no premiss undischarged before (1) and since (1) itself is the only premiss undischarged at (1) this line on our new interpretation has the correct meaning that the subsidiary deduction *from* (1) *to* (7) yields (8).

Example x

All propositions which have factual content are empirical hypotheses. Every empirical hypothesis provides a rule for the anticipation of experience. An empirical hypothesis provides a rule for the anticipation of experience if and only if it is relevant to experience. The principle of verifiability is true if and only if every proposition which has factual content is relevant to experience. Therefore the principle of verifiability is true.

(Adapted from A. J. Ayer: *Language, Truth and Logic* (1st edition, pp. 30–31).)

With the following abbreviations: $F(\ )$ for ( ) *is a proposition which has factual content*, $H(\ )$ for ( ) *is an empirical hypothesis*, $P(\ )$ for ( ) *provides a rule for the anticipation of experience*, $R(\ )$ for ( ) *is relevant to experience*, t for *the principle of verifiability is true*, the argument may be expressed as follows:

$$(\forall x)[Fx \supset Hx],\ (\forall x)[Hx \supset Px],\ (\forall x)[Hx \supset [Px \equiv Rx]],\ t \equiv (\forall x)[Fx \supset Rx];\ \text{therefore } t.$$

It may be shown to be valid by means of this deduction.

Pr × 1
(1) $(\forall x)[Fx \supset Hx]$

Pr × 2
(2) $(\forall x)[Hx \supset Px]$

Pr × 3
(3) $(\forall x)[Hx \supset [Px \equiv Rx]]$

Pr × 4
(4) $t \equiv (\forall x)[Fx \supset Rx]$
1 UI × 5
(5) $Fa \supset Ha$
2 UI × 6
(6) $Ha \supset Pa$
3 UI × 7
(7) $Ha \supset [Pa \equiv Ra]$

Pr × 8
(8) Ha
7, 8 MP × 9
(9) $Pa \equiv Ra$
9 TF × 10
(10) $Pa \supset Ra$
6, 10, 8 TF × 11
(11) Ra

(8⊢11) × 12
(12) $Ha \supset Ra$
5, 12 HS × 13
(13) $Fa \supset Ra$
13 UG × 14
(14) $(\forall x)[Fx \supset Rx]$
4, 14 TF × 15
(15) t

As may be seen from the list of abbreviations, there is no proper name in any of the premisses undischarged before (14). *a* may therefore be any proper name.

Logical principles	Fallacies
(27) $(\forall x)[Fx . p] \equiv [(\forall x)Fx . p]$	
(28) $(\exists x)[Fx . p] \equiv [(\exists x)Fx . p]$	
(29) $(\forall x)[Fx \vee p] \equiv [(\forall x)Fx \vee p]$	
(30) $(\exists x)[Fx \vee p] \equiv [(\exists x)Fx \vee p]$	
(31) $(\forall x)[p \supset Fx] \equiv [p \supset (\forall x)Fx]$	
(32) $(\exists x)[p \supset Fx] \equiv [p \supset (\exists x)Fx]$	
(33) $(\forall x)[Fx \supset p] \equiv [(\exists x)Fx \supset p]$	
(34) $(\exists x)[Fx \supset p] \equiv [(\forall x)Fx \supset p]$	
(35) $(\forall x)[Fx \supset p] \supset [(\forall x)Fx \supset p]$	(35.1) Converse of 35
(36) $[(\exists x)Fx \supset p] \supset (\exists x)[Fx \supset p]$	(36.1) Converse of 36
(37) $(\forall x)[Fx \equiv p] \supset [(\forall x)Fx \equiv p]$	(37.1) Converse of 37

3.3. Principles of single quantification; deductions

The symbol p in (27) to (37) represents a formula which is a statement and as such does not contain a free variable, x or any other. For example, p might be the statement *There will be a holiday* and Fx might mean *x will pass the examination*. Then the left-hand side of (33), for example, will mean (a) *Of everyone it is true that if he passes the examination there will be a holiday* and the right-hand side will mean (b) *if someone passes the examination there will be a holiday*. (33) then implies that (a) and (b) are equivalent, which is indeed only what one should expect in view of the fact that both (a) and (b) might quite naturally be expressed in the form *if anyone passes the examination there will be a holiday*. The logical truth of (33) is proved formally in example (xi).

The meaning of the left-hand side of (34) may not be obvious at first sight. The reader should, however, be able to interpret it in the light of the remarks made with reference to (23) on page 74; he is advised to try to do this and also to convince himself, if possible both by means of a formal deduction and intuitively, of the logical truth of the rather surprising right-left implication in (34), i.e. of the implication $[(\forall x)Fx \supset p] \supset (\exists x)[Fx \supset p]$.

Our concluding list in this section contains four principles all of which are of some interest. The first two (38) and (39) in effect assert jointly that where there is only single quantification the choice of letter for the pronoun variable has no significance. The other two assert in effect that the result of putting an x-quantifier in front of a statement rather than a pronominal clause and so in front of a formula which does not contain free x, is simply a statement which is equivalent to the original one.

Logical principles	Fallacies
(38) $(\forall x)Fx \equiv (\forall y)Fy$	
(39) $(\exists x)Fx \equiv (\exists y)Fy$	
(40) $(\forall x)p \equiv p$	
(41) $(\exists x)p \equiv p$	

3.3. Principles of single quantification; deductions

Example xi

Pr × 1	Pr × 8
(1) $(\forall x)[Fx \supset p]$	(8) $(\exists x)Fx \supset p$
Pr × 2	Pr × 9
(2) $(\exists x)Fx$	(9) $\sim(\forall x)[Fx \supset p]$
2 EI × 3	9 int (XV) × 10
(3) Fd	(10) $(\exists x) \sim [Fx \supset p]$
1 UI × 4	10 EI × 11
(4) $Fd \supset p$	(11) $\sim [Fd_1 \supset p]$
4, 3 MP × 5	11 TF × 12, 13
(5) p	(12) Fd_1
	(13) $\sim p$
(2⊢3) × 6	12 EG × 14
(6) $(\exists x)Fx \supset p$	(14) $(\exists x)Fx$
	8, 14, MP × 15
(1⊢6) × 7	(15) p
(7) $(\forall x)[Fx \supset p] \supset [(\exists x)Fx \supset p]$	13, 15, TF × 16
	(16) $(\forall x)[Fx \supset p]$
	(9⊢16) × 17
	(17) $\sim(\forall x)[Fx \supset p] \supset (\forall x)[Fx \supset p]$
	17 TF × 18
	(18) $(\forall x)[Fx \supset p]$
	(8⊢18) × 19
	(19) $[(\exists x)Fx \supset p] \supset (\forall x)[Fx \supset p]$
	7, 19, TF × 20
	(20) $(\forall x)[Fx \supset p] \equiv [(\exists x)Fx \supset p]$

3.3. Principles of single quantification; deductions

A deduction establishing the left-right implication in (39) is given in example (xii). The right-left implication could of course be established by the same deduction with x and y interchanged throughout. In example (xiii) we have a deduction establishing (40).

Example xii

Pr × 1
(1) $(\exists x)Fx$
1 EI × 2
(2) Fd
2 EG × 3
(3) $(\exists y)Fy$

(1⊢3) × 4
(4) $(\exists x)Fx \supset (\exists y)Fy$

Example xiii

Pr × 1
(1) $(\forall x)p$
1 UI × 2
(2) p

(1⊢2) × 3
(3) $(\forall x)p \supset p$

Pr × 4
(4) p
4 UG a/x × 5
(5) $(\forall x)p$

(4⊢5) × 6
(6) $p \supset (\forall x)p$
3, 6 TF × 7
(7) $(\forall x)p \equiv p$

a must be a proper name not contained in p.

There are two points to be commented on here. First, since p, being a statement, does not contain any free variable and so no free x the application of UI to (1) gives p since p is the result of substituting a for x at every free occurrence in p. Secondly, since the a mentioned in the justification line for (5) is a proper name not contained in p, the application of UG to (4) there specified results in $(\forall x)p$, since the substitution of x for a at every occurrence in p, where a has no occurrence in p, results in p itself.

4. Multiple quantification; introduction. A formula may be said to be multiply quantified when it contains at least one instance of one quantifier occurring within the scope of another. For example $(\forall x)[Fx.(\exists y)Gxy]$ is multiply quantified since $(\exists y)$ is within the scope of $(\forall x)$; on the other hand, $(\forall x)Fx.(\exists y)Gyy$ is not multiply quantified since, although it contains two quantifiers, neither is within the scope of the other. We have already had several examples of multiply quantified formulae.

In what sort of circumstances are multiply quantified statements appropriate and what is their significance? We may obtain some light on these questions by considering a simple illustration. Suppose that we have a limited universe of discourse consisting of Tom, Dick and Harry who were once on military service together. Let us write down a list of statements which might be made about these men and their army numbers:

(1) Tom remembers his own number.
(2) Dick remembers Tom's number.
(3) Harry remembers Tom's number.
(4) Tom remembers Dick's number.
(5) Dick remembers his own number.
(6) Harry remembers Dick's number.
(7) Tom remembers Harry's number.
(8) Dick remembers Harry's number.
(9) Harry remembers his own number.

Given that any one of these statements is true the multiply quantified statement $(\exists y)[(\exists x)[y$ *remembers x's number*$]]$

must be true. Suppose for example that (4) is true. We may then assert:

(4′) (Dick x)[Tom remembers x's number].

From (4′) in turn we may derive:

(10) $(\exists x)$[Tom remembers x's number].

But (10) of course is equivalent to

(10′) (Tom y)[$(\exists x)$[y remembers x's number]];

from which follows immediately:

(11) $(\exists y)$[$(\exists x)$[y remembers x's number]].

Suppose now that (1), (2) and (6) are all true. Then the equivalent statements:

(1′) (Tom x)[Tom remembers x's number].
(2′) (Tom x)[Dick remembers x's number]
(6′) (Dick x)[Harry remembers x's number]

are all true. From these three statements it is obvious that we may derive respectively:

(12) $(\exists x)$[Tom remembers x's number]
(13) $(\exists x)$[Dick remembers x's number]

and

(14) $(\exists x)$[Harry remembers x's number].

Given (12), (13) and (14) the equivalent statements:

(12′) (Tom y)[$(\exists x)$[y remembers x's number]]
(13′) (Dick y)[$(\exists x)$[y remembers x's number]]
(14′) (Harry y)[$(\exists x)$[y remembers x's number]]

must be true, and, since Tom, Dick and Harry are by hypothesis the only members of the universe of discourse, the matrix [$(\exists x)$[y *remembers* x*'s number*]] is true of every individual as y; hence we may assert:

(15) $(\forall y)$[$(\exists x)$[y remembers x's number]].

3.4. Multiple quantification; introduction

Suppose again that (1), (2) and (3) are true statements. The equivalent statements:

(1″) (Tom y)[y remembers Tom's number]
(2″) (Dick y)[y remembers Tom's number]
(3″) (Harry y)[y remembers Tom's number]

must also be true.

It follows that the pronominal clause *y remembers Tom's number* is true of every individual as y so that we may assert:

(16) $(\forall y)$[y remembers Tom's number].

If (16) is true the equivalent statement:

(16′) (Tom x)[$(\forall y)$[y remembers x's number]]

is true also and in view of this we may assert:

(17) $(\exists x)$[$(\forall y)$[y remembers x's number]].

Lastly we can easily see that in the same way that we obtained (16) from (1), (2) and (3), so we obtain

(18′) (Dick x)[$(\forall y)$[y remembers x's number]]
from (4), (5) and (6)

and

(19′) (Harry x)[$(\forall y)$[y remembers x's number]]
from (7), (8) and (9).

It is clear that if all the statements (1) to (9) are true then (16′), (18′) and (19′) are all true. If these statements are true the pronominal clause $(\forall y)$[*y remembers x's number*] is true of every individual as x and we may assert:

(20) $(\forall x)$[$(\forall y)$[y remembers x's number]].

Disregarding statements containing singular quantifiers we have now four examples of multiply quantified statements, all with the same matrix:

(11) $(\exists y)[(\exists x)[y$ remembers x's number]]
(15) $(\forall y)[(\exists x)[y$ remembers x's number]]
(17) $(\exists x)[(\forall y)[y$ remembers x's number]]
(20) $(\forall x)[(\forall y)[y$ remembers x's number]]

Each of these statements has been based on a different subset of the original nine. Some of them could have been arrived at from alternative subsets; e.g. (11) could have been obtained from any statement other than (4). It may be noted, however, that although (15) could be obtained from any set of statements from which (17) could be obtained the converse is not true; indeed (17) could not be obtained from the set of three statements (1), (2), (6), from which we did obtain (15). As we shall prove later in a general way[1] (15) must be true if (17) is true, but the converse does not hold. A study of these four statements and of the different ways which they and other similar ones may be obtained should lead to a reasonably sound understanding of the significance of multiply quantified statements in general. It may help if we give informal renderings:

(11) It is true of someone that he remembers someone's number;
(15) It is true of everyone that he remembers someone's number;
(17) It is true of someone that everyone remembers his number;
(20) It is true of everyone that everyone remembers his number.

It is usual to omit brackets where these are not necessary to prevent ambiguity. Thus our four statements could be written:

[1] See example (xvi) in the present chapter, and example 5 in chapter 4, section 6.

$(\exists y)(\exists x)$ *y* remembers *x*'s number;
$(\forall y)(\exists x)$ *y* remembers *x*'s number;
$(\exists x)(\forall y)$ *y* remembers *x*'s number;
$(\forall x)(\forall y)$ *y* remembers *x*'s number.

The reader may test his understanding by considering the four statements which begin with the same pair of quantifiers as these four but have the matrix: *x remembers y's number*. From what subset could each of these statements be derived? What interrelationships of meaning exist between the two groups of statements?

No new principles are involved in the consideration of arguments involving multiple quantification and these are shown to be valid or invalid as the case may be by exactly the same methods as are used for arguments involving only single quantification. In particular no additions are needed to our rules of deduction, and we give now an example of a deduction in accordance with our existing rules which establishes the validity of the argument form:

$(\exists x)[Fx.(\forall y)Gxy]$; *therefore* $(\forall y)(\exists x)[Fx.Gxy]$.

Example xiv

	Pr × 1		4 UI × 5
(1)	$(\exists x)[Fx.(\forall y)Gxy]$	(5)	Gda
	1 EI × 2		3, 5 Conj × 6
(2)	$Fd.(\forall y)Gdy$	(6)	$Fd.Gda$
	2 Simp × 3, 4		6 EG × 7
(3)	Fd	(7)	$(\exists x)[Fx.Gxa]$
			7 UG × 8
(4)	$(\forall y)Gdy$	(8)	$(\forall y)(\exists x)[Fx.Gxy]$

a must be a proper name not contained in *F*( ) or *G*(1)(2)

It can be seen in this example that each single step differs in no significant way from many steps which we have already had in examples i to xiii.

5. Expressing multiply general statements in quantificational form. However, although deductions involving multiple quantification are not as a rule found to present any special difficulty it cannot be denied that many students do find great difficulty with a related but equally important operation, namely that of rendering ordinary discourse general arguments into quantificational form, where multiple quantification is involved. Accordingly we shall devote a little time to outlining and explaining procedures by means of which it is hoped that the difficulty of this operation may at least be reduced.

We shall set forth to begin with a procedure for dealing with a certain limited class of general statements which we may call the class A. A general statement belongs to class A if it is governed by one generalizer which has not more than one other generalizer within its scope. Let ( ) *is f*, ( ) *is g* represent any one-place predicates of ordinary discourse and let (*1*) *is r to* (*2*) represent any two-place predicate; for example ( ) *is f* might be (   ) *is wise* or (   ) *speaks French* and (*1*) *is r to* (*2*) might be (*1*) *admires* (*2*). Then statements of the following forms are some of those which belong to class A:

(i) Everything is *f*,
(ii) Something is *f*,
(iii) Everything which is *f* is *g*,
(iv) Something which is *f* is *g*,
(v) Everything which is *f* is *r* to something which is *g*.

It is hoped that as a result of our discussion in chapter one readers have now no difficulty in expressing (i) to (iv) in quantificational form, and the procedure which we are going to give is intended primarily to help with doubly general statements such as (v). However, it can be used for the basic singly general statements also, whether qualified or unqualified.

3.5. Expressing multiply general statements

In any procedure for expressing a multiply general statement in quantificational form the first important step must be to decide which is dominant among the generalizers; to do this we have to be able to decide which of any two connected generalizers is dominant. How can one tell which of two generalizers is dominant? If both are universal or both are particular the question is of no logical importance; one may make it a rule to treat the one which occurs first as dominant. If one generalizer is universal and the other particular, one may use a criterion which we will explain in relation to an example. Suppose that in a certain context we find the statement:

(G) Someone is liked by everyone.

Which of the two generalizers is dominant, *someone* or *everyone*? Now (G) must be equivalent to one or other of the two statements:

(P) Some statement of the form *a is liked by everyone* is true;
(U) Every statement of the form *Someone is liked by a* is true.

Which of these two statements represents what is intended by (G) has to be determined by the context. If (P) is intended then the particular generalizer *someone* is dominant. If (U) is intended then the universal generalizer *everyone* is dominant. Usually the first generalizer is dominant but this cannot be taken as an invariable rule: one can imagine contexts for (G) in which the universal generalizer would be dominant. In our examples, however, in the absence of contrary indications, we will take the first generalizer to be dominant.

Procedure for statements of class A (Simple singly or doubly general statements—see above, page 87)

(i) Write down the statement and number it (1).

(ii) Write down an equivalent statement with the dominant generalizer at the beginning in the standard form:

Every individual it is such that

or

Some individual it is such that.

Number this statement (1a).

(iii) Replace the *it* introduced in (ii) at every occurrence in (1a) by the same variable **x**. **x** must be a variable not previously used. Number the resulting statement (1b).

In (ii) and (iii) *it* stands for whatever pronoun is appropriate.

(iv) Replace the initial quantifier in (1b) by the appropriate symbol and enclose the remaining part of (1b) in brackets. This gives a statement in quantifier-matrix (Q–M) form consisting of a symbolic quantifier followed by a matrix. Number this statement (1′).

(v) Write down the matrix of (1′) separately and number it (2).

(vi) Express (2) in truth-functional form, abbreviating any non-generalized component. Number resulting formula (2′).

(vii) If (2′) contains no generalized component (as will be the case if (1) is just singly general) renumber (2′) as (2*). In (1′) replace (2) by (2*) and number the resulting formula (1*). (1*) is the required formula.

If (2′) contains one generalized component (as will be the case if (1) is doubly general), go to instruction (viii).

(viii) Write down the generalized component separately and number it (3).

(ix) Repeat the procedure stated in (ii) to (vi) for (3) instead of for (1); i.e. begin by increasing the number in each of (1), (1a), (1b), (1′), (2), (2′) by 2, so that these symbols in the instructions (ii) to (vi) read respectively: 3, 3a, 3b, 3′, 4, 4′.

(x) Renumber (4′) as (4*.)

(xi) In (3′) replace (4) by (4*) and number resulting formula (3*).

(xii) In (2′) replace (3) by (3*) and number resulting formula (2*).

(xiii) In (1′) replace (2) by (2*) and number resulting formula (1*). (1*) is the required formula.

We give two examples of the application of this procedure. In examples 1 and 2 the procedure is used to express in quantificational form respectively: *Every new member is known to some officer*, *At least one f is r to some g*. In example 1 at each line the instruction which is being followed is indicated at the left-hand side.

Example 1

i	Every new member is known to some officer	1	
ii	Every individual he is such that if he is a new member he is known to some officer	1a	
iii	Every individual x is such that if x is a new member x is known to some officer	1b	
iv	$(\forall x)$[if x is a new member x is known to some officer]	1′	
v	[if x is a new member x is known to some officer]		2
vi	[$Nx \supset x$ is known to some officer]		2′
vii	*2′ does contain a generalized component*		
viii	x is known to some officer		3

ix ii Some individual he is such that he is an officer and x is known to him — 3a

iii Some individual y is such that y is an officer and x is known to y — 3b

iv $(\exists y)$[y is an officer and x is known to y] — 3′

v [y is an officer and x is known to y] — 4

vi $[Oy.Kxy]$ — 4′

x $[Oy.Kxy]$ — 4*

xi $(\exists y)[Oy.Kxy]$ — 3*

xii $[Nx \supset (\exists y)[Oy.Kxy]\,]$ — 2*

xiii $(\forall x)[Nx \supset (\exists y)[Oy.Kxy]\,]$ — 1*

Example 2

At least one f is r to some g — 1

Some individual it is such that it is an f and it is r to some g — 1a

Some individual x is such that x is an f and x is r to some g — 1b

$(\exists x)$[x is an f and x is r to some g] — 1′

[x is an f and x is r to some g] — 2

[$Fx.x$ is r to some g] — 2′

x is r to some g — 3

Some individual it is such that it is a g and x is r to it — 3a

Some individual y is such that y is a g and x is r to y — 3b

$(\exists y)$[y is a g and x is r to y] — 3′

[y is a g and x is r to y] — 4

$[Gy.Rxy]$ — 4′, 4*

$(\exists y)[Gy.Rxy]$ — 3*

$[Fx.(\exists y)[Gy.Rxy]\,]$ — 2*

$(\exists x)[Fx.(\exists y)[Gy.Rxy]\,]$ — 1*

3.5. *Expressing multiply general statements*

In example 1 we have followed the instructions exactly and have allowed a separate line for each instruction. In example 2 we have included an obvious abbreviation of one line. The student is likely after a little practice to abbreviate the procedure in ways which suit himself.

We show below one possible shortened version of the procedure as applied to example 1.

Every new member is known to some officer (1)

$(\forall x)|$ [if x is a new member x is known to some officer] (1′)|(2)

$[Nx \supset |\, x$ is known to some officer] (2′)|(3)

$(\exists y|$ [y is an officer and x is known to y] (3′)|(4)

$(\forall x)|\,[Nx \supset |(\exists y)|\,[Oy.Kxy]\,]$ (1*)|(2*)|(3*)|(4′)(4*)

The last line here is arrived at as follows. In accordance with (vi), repeated under (ix), we obtained $[Oy.Kxy]$ as the truth-functional expression of the matrix (4) of (3′) and number it (4′). In accordance with (x) we renumber it (4*). We then obtain successively (3*), (2*) and (1*). The vertical lines are of course only for guidance in the abbreviated procedure. It will be seen that in practice the numbering can also be dispensed with.

It is desirable, and perfectly possible, to attain a facility such that the quantificational form of any class A statement can be written down at once, without any intermediate steps.

We now set forth a completely general procedure which is applicable to any relevant statement whether of class A or not.

General procedure. In the instructions M means the left-hand margin; m in any instruction means the number that is in M when that instruction is being followed.

Instructions

(i) Write -2 in M.

(ii) Write down the formula to be expressed in quantificational form and number it (0).

(iii) Increase the number in M by 2.

(iv) Express (m) truth-functionally and abbreviate any non-generalized truth-functional components. Number the resulting formula (m').

(v) If (m') has no unanalysed general component (UGC) renumber it (m^*) and go to instruction (ix). If (m') has a UGC, go to instruction (vi).

(vi) Delete all numbers $>m$ by which any earlier formulae are numbered (including primed and starred numbers); write down the first UGC in m' separately and number it ($m+1$).

(vii) Express ($m+1$) in quantifier-matrix (Q-M) form, using a pronoun variable not previously used and number result $(m+1)'$.

(viii) Write down matrix of $(m+1)'$ separately and number it ($m+2$). Return to instruction (iii).

(ix) If $m=0$, (m^*) is the required formula. The procedure ends.
If $m>0$, go to instruction (x).

(x) In $(m-1)'$ replace (m) by (m^*).
If resulting formula has a UGC transfer number $(m-1)'$ to it and go to instruction (xi).
If resulting formula has no UGC number it $(m-1)^*$ and go to instruction (xii).

(xi) Reduce number in M by 1. Go to instruction vi.

(xii) Reduce number in M by 1. Go to instruction ix.

Example 3

The general procedure is used to express in quantificational form the formula: *p and every f, if q, is r to some g and s to every h. p* and *q* represent unanalysed statements which are to be regarded as elements.

In the left-hand margin at any line the arabic numeral (if there is any) represents the number in *M*, the small roman numeral shows the instruction which is being followed and the minute numeral is the serial number of the line, inserted for ease of reference in subsequent explanatory remarks.

M	Instr.	Line	Expression				
−2	i	1	—				
	ii	2	*p* and every *f*, if *q*, is *r* to some *g* and *s* to every *h*	0			
0	iii	3	—				
	iv	4	*p*. every *f*, if *q*, is *r* to some *g* and *s* to every *h*	0′			
	v	5	—				
	vi	6	every *f*, if *q*, is *r* to some *g* and *s* to every *h*		1		
	vii	7	(∀*x*)[if *x* is *f* then if *q x* is *r* to some *g* and *s* to every *h*]		1′		
	viii	8	[if *x* is *f* then if *q x* is *r* to some *g* and *s* to every *h*]			2	
2	iii	9	—				
	iv	10	$[Fx \supset [q \supset [x$ is *r* to some *g*.*x* is *s* to every *h*]]]			2′	
	v	11	—				
	vi	12	*x* is *r* to some *g*			3	
	vii	13	(∃*y*)[*y* is a *g* and *x* is *r* to *y*]			3′	
	viii	14	[*y* is a *g* and *x* is *r* to *y*]				4
4	iii	15	—				
	iv	16	[*Gy*.*Rxy*]				4′
	v	17	[*Gy*.*Rxy*]				4*
	ix	18	—				
	x	19	(∃*y*)[*Gy*.*Rxy*]			3*	
3	xii	20	—				
	ix	21	—				

in quantificational form

x 22 $[Fx \supset [q \supset [(\exists y)[Gy . Rxy]$
$. x$ is s to every $h]]]$ 2′

2 xi 23 —

vi 24 x is s to every h 3

vii 25 $(\forall z)$[if z is a h x is s to z] 3′

viii 26 [if z is a h x is s to z] 4

4 iii 27 —

iv 28 $[Hz \supset Sxz]$ 4′

v 29 $[Hz \supset Sxz]$ 4*

ix 30 —

x 31 $(\forall z)[Hz \supset Sxz]$ 3*

3 xii 32 —

ix 33 —

x 34 $[Fx \supset [q \supset [(\exists y)[Gy . Rxy]$
$. (\forall z)[Hz \supset Sxz]]]]$ 2*

2 xii 35 —

ix 36 —

x 37 $(\forall x)[Fx \supset [q \supset [(\exists y)[Gy . Rxy]$
$. (\forall z)[Hz \supset Sxz]]]]$ 1*

1 xii 38 —

ix 39 —

x 40 $p . (\forall x)[Fx \supset [q \supset [(\exists y)[Gy . Rxy]$
$. (\forall z)[Hz \supset Sxz]]]]$ 0*

0 xii 41 —

ix 42 (0*) is required formula.

The number (2′) against the formula in line 10 is cancelled when this number is transferred to the formula in line 22. The numbers >2 against formulae in lines 12 to 19 are cancelled by instruction vi at line 24.

By short-cuts of the kind used in connexion with example 2 the number of lines needed for example 3 can be greatly reduced.

p and every f, if q, is r to some g
and s to every h (0)

$p . |$every f, if q, is r to some g
and s to every h (0′) |(1)

$(\forall x)\vert$ [if x is f then if q x is r to some g and s to every h]	(1′) |(2)
$[Fx \supset [q \supset [\vert x$ is r to some $g\vert . x$ is r to every $h]]]$	(2′)|(3)
$(\exists y)\vert [y$ is a $g . x$ is r to $y]$	(3′) |(4)
$(\exists y)\vert [Gy . Rxy]$	(3*)|(4′)(4*)
$[Fx \supset [q \supset [(\exists y)[Gy . Rxy] . \vert x$ is s to every $h]]]$	(2′) |(3)
$(\forall z)\vert$ [if z is a h x is s to z]	(3′) |(4)
$p . \vert(\forall x)\vert [Fx \supset [q \supset [(\exists y)[Gy . Rxy] . \vert (\forall z)\vert [Hz \supset Sxz]]]]$	(0*)|(1*)|(2*)|(3*)|(4′)(4*)

6. Principles of multiple quantification; deductions.

Logical principles	Fallacies
(42) $(\forall x)(\forall y)Fxy \supset (\forall x)(\exists y)Fxy$	
(43) $(\forall x)(\forall y)Fxy \supset (\exists x)(\exists y)Fxy$	

Example xv

The logical truth of (42) is proved by the following deduction.

Pr × 1
(1) $(\forall x)(\forall y)Fxy$
1 UI × 2
(2) $(\forall y)Fay$
2 UI × 3
(3) Faa_1
3 EG × 4
(4) $(\exists y)Fay$
4 UG × 5
(5) $(\forall x)(\exists y)Fxy$
(1⊢5) × 6
(6) $(\forall x)(\forall y)Fxy \supset (\forall x)(\exists y)Fxy$

a must be a proper name not contained in F(1)(2).

Logical principles	Fallacies
(44) $(\forall x)(\forall y)Fxy \equiv (\forall y)(\forall x)Fxy$	
(45) $(\exists x)(\exists y)Fxy \equiv (\exists y)(\exists x)Fxy$	
(46) $(\exists x)(\forall y)Fxy \supset (\forall y)(\exists x)Fxy$	(46.1) $(\forall y)(\exists x)Fxy \supset (\exists x)(\forall y)Fxy$

From (44) to (46.1) we see that if two quantifiers occur together in a formula F then if they are of the same type (either both universal or both particular) their order makes no difference; but on the other hand if one is universal and the other particular a change in order may affect the truth-value of F.

Example xvi

(46) is proved by the following deduction.

Pr × 1	3 EG × 4
(1) $(\exists x)(\forall y)Fxy$	(4) $(\exists x)Fxa$
1 EI × 2	4 UG × 5
(2) $(\forall y)Fdy$	(5) $(\forall y)(\exists x)Fxy$
2 UI × 3	(1⊢5) × 6
(3) Fda	(6) $(\exists x)(\forall y)Fxy \supset (\forall y)(\exists x)Fxy$

a must be a proper name not contained in F(1)(2).

Some might be tempted to think that the converse of (46), i.e. (46.1), may be proved by the sequence which we now give.

	Pr × 1			3 UG × 4
(1)	$(\forall y)(\exists x)Fxy$	WRONG {	(4)	$(\forall y)Fdy$
	1 UI × 2			4 EG × 5
(2)	$(\exists x)Fxa$		(5)	$(\exists x)(\forall y)Fxy$
	2 EI × 3			$(1 \vdash 5) \times 6$
(3)	Fda		(6)	$(\forall y)(\exists x)Fx \supset (\exists x)(\forall y)Fxy$

However, this sequence is not a proper deduction; since line (3) *Fda* contains a dummy name the rule UG has been applied to it illegitimately. It should be understood of course that the fact that this particular way of attempting to deduce (46.1) is not allowable does not by itself entitle us to conclude that (46.1) is not logically true; for there might well be some other and correct way of deducing it which we have not been able to think of. However, (46.1) is in fact fallacious and this will be properly proved in the next chapter.[1]

There are many principles involving multiple quantification which are in a sense simple extensions of principles involving only single quantification. We give a few examples in the next list.

Logical principles	Fallacies
(47) $(\forall x)(\forall y)[Fxy \supset Gxy]$, Faa_1; $\therefore Gaa_1$	(47.1) $(\forall x)(\forall y)[Fxy \supset Gxy]$, Gaa_1; $\therefore Faa_1$
(48) $(\forall x)(\forall y)[Fxy \supset Gxy] \supset [(\forall x)(\forall y)Fxy \supset (\forall x)(\forall y)Gxy]$	(48.1) Converse of 48
(49) $(\forall x)(\forall y)[Gxy \supset Hxy]$, $(\exists x)(\exists y)[Fxy . Gxy]$; $\therefore (\exists x)(\exists y)[Fxy . Hxy]$	

[1] Section 6, example 5.

Example xvii

The validity of (49) is established in this deduction.

	Pr × 1		5 UI × 6
(1)	$(\forall x)(\forall y)[Gxy \supset Hxy]$	(6)	$Gdd_1 \supset Hdd_1$
	Pr × 2		6, 4 TF × 7
(2)	$(\exists x)(\exists y)[Fxy . Gxy]$	(7)	$Fdd_1 . Hdd_1$
	2 EI × 3		7 EG × 8
(3)	$(\exists y)Fdy . Gdy$	(8)	$(\exists y)[Fdy . Hdy]$
	3 EI × 4		8 EG × 9
(4)	$Fdd_1 . Gdd_1$	(9)	$(\exists x)(\exists y)[Fxy . Hxy]$
	1 UI × 5[1]		
(5)	$(\forall y)[Gdy \supset Hdy]$		

The next list contains two principles (50) and (51) which perhaps belong more distinctively to multiple quantification though they are related in a fairly obvious way to (31) and (27) respectively.

Logical principles	Fallacies
(50) $(\forall x)(\forall y)[Fx \supset Gxy] \equiv (\forall x)[Fx \supset (\forall y)Gxy]$	
(51) $(\exists x)(\forall y)[Fx . Gxy] \equiv (\exists x)[Fx . (\forall y)Gxy]$	
(52) $(\exists y)(\forall x)[Fx . Gxy] \supset (\forall x)[Fx . (\exists y)Gxy]$	(52.1) Converse of 52

The right-left implication in (51), though logically true presents a rather difficult problem of deduction, since though we may obtain in an obvious way, e.g. $Fd . Gda$ we cannot owing to the presence of *d* go on to restore the universal quantifier $(\forall y)$. We show in example xviii a method whereby the deduction may nevertheless be done.

Example xviii

Pr × 1
(1) $(\exists x)[Fx.(\forall y)Gxy]$

1 EI × 2
(2) $Fd.(\forall y)Gdy$

Pr × 3
(3) $Fa.(\forall y)Gay$

3 TF × 4
(4) $(\forall y)Gay$
4 UI × 5
(5) Gaa_1

3, 5 TF × 6
(6) $Fa.Gaa_1$

6 UG × 7
(7) $(\forall y)[Fa.Gay]$

(3⊢7) × 8
(8) $[Fa.(\forall y)Gay] \supset (\forall y)[Fa.Gay]$
8 UG × 9
(9) $(\forall x)[[Fx.(\forall y)Gxy] \supset (\forall y)[Fx.Gxy]]$

9 UI × 10
(10) $[Fd.(\forall y)Gdy] \supset (\forall y)[Fd.Gdy]$
10, 2 MP × 11
(11) $(\forall y)[Fd.Gdy]$
11 EG × 12
(12) $(\exists x)(\forall y)[Fx.Gxy]$

(1⊢12) × 13
(13) $(\exists x)[Fx.(\forall y)Gxy] \supset (\exists x)(\forall y)[Fx.Gxy]$

a and a_1 must be proper names not contained in $F()$ or $G(1)(2)$.

In our final list we give some principles which are useful extensions of XII to XV in our list of equivalences.

Logical principles	Fallacies
(53) $(\forall x)(\forall y)Fxy \equiv \sim(\exists x)(\exists y)\sim Fxy$	
(54) $(\exists x)(\exists y)Fxy \equiv \sim(\forall x)(\forall y)\sim Fxy$	
(55) $(\forall x)(\forall y)\sim Fxy \equiv \sim(\exists x)(\exists y)Fxy$	
(56) $(\exists x)(\exists y)\sim Fxy \equiv \sim(\forall x)(\forall y)Fxy$	

Example xix

In the following deduction the left-right implication in (53) is shown to be logically true.

Pr × 1
(1) $(\forall x)(\forall y)Fxy$
1 UI × 2
(2) $(\forall y)Fay$
2 UI × 3
(3) Faa_1

3 TF × 4
(4) $\sim\sim Faa_1$

4 UG × 5
(5) $(\forall y)\sim\sim Fay$

5 int (XIV) × 6
(6) $\sim(\exists y)\sim Fay$
6 UG × 7
(7) $(\forall x)\sim(\exists y)\sim Fxy$
7 int (XIV) × 8
(8) $\sim(\exists x)(\exists y)\sim Fxy$

(1⊢8) × 9
(9) $(\forall x)(\forall y)Fxy \supset \sim(\exists x)(\exists y)\sim Fxy$

a and a_1 must be proper names not contained in *F*(*1*)(*2*).

We conclude this section with some miscellaneous examples.

Example xx

No member of the first-year class has read any work of Kant.
The Critique of Pure Reason is by Kant and Smith has read it.
Therefore Smith is not a member of the first-year class.

In quantificational form this argument becomes:

$(\forall x)[Fx \supset (\forall y)[Ky \supset \sim Rxy]]$;
$Ka.Ra_1a$; therefore $\sim Fa_1$,

where the abbreviations: *F*( ) for (   ) *is a member of the first-year class*, *K*( ) for (   ) *is a work of Kant*, *R*(*1*)(*2*) for (*1*) *has read* (*2*), *a* for *The Critique of Pure*

Reason, and a_1 for *Smith*, are used. It is shown to be valid by the following deduction:

Pr × 1
(1) $(\forall x)[Fx \supset (\forall y)[Ky \supset \sim Rxy]]$

Pr × 2
(2) $Ka.Ra_1a$

1 UI × 3
(3) $Fa_1 \supset (\forall y)[Ky \supset \sim Ra_1y]$

Pr × 4
(4) Fa_1

2, 4 MP × 5
(5) $(\forall y)[Ky \supset \sim Ra_1y]$

5 UI × 6
(6) $Ka \supset \sim Ra_1a$

2, 6 TF × 7, 8
(7) Ra_1a

(8) $\sim Ra_1a$

7, 8 TF × 9
(9) $\sim Fa_1$

(4⊢9) × 10
(10) $Fa_1 \supset \sim Fa_1$

10 TF × 11
(11) $\sim Fa_1$

Example xxi

> Anyone who supported this appeal is either a sentimentalist or a friend of Mr. Jones. Mr. Jones is a member of the committee. No one who lives here is a sentimentalist. Therefore if anyone who lives here supported the appeal he is a friend of a member of the committee.

With obvious abbreviations (including *a* for *Mr. Jones*) this argument is expressed in quantificational form thus:

$(\forall x)[Ax \supset [Sx \vee Fxa]]$;
Ca;
$(\forall x)[Lx \supset \sim Sx]$; therefore $(\forall x)[Lx \supset [Ax \supset (\exists y)[Cy.Fxy]]]$.

In the following deduction it is shown to be valid.

Pr × 1

(1) $(\forall x)[Ax \supset [Sx \vee Fxa]]$

Pr × 2

(2) Ca

Pr × 3

(3) $(\forall x)[Lx \supset \sim Sx]$

1 UI × 4

(4) $Aa_1 \supset [Sa_1 \vee Fa_1 a]$

3 UI × 5

(5) $La_1 \supset \sim Sa_1$

Pr × 6

(6) La_1

Pr × 7

(7) Aa_1

4, 5, 6, 7 TF × 8

(8) $Fa_1 a$

2, 8 conj × 9

(9) $Ca . Fa_1 a$

9 EG × 10

(10) $(\exists y)[Cy . Fa_1 y]$

$(7 \vdash 10)$ × 11

(11) $Aa_1 \supset (\exists y)[Cy . Fa_1 y]$

$(6 \vdash 11)$ × 12

(12) $La_1 \supset [Aa_1 \supset (\exists y)[Cy . Fa_1 y]]$

12 UG × 13

(13) $(\forall x)[Lx \supset [Ax \supset (\exists y)[Cy . Fxy]]]$

Since a (=Mr. Jones) is the only proper name contained in the premisses ((1), (2), (3)) which are undischarged when *UG* is applied to obtain line (13), it is sufficient to say here that a_1 must be a proper name other than a.

7. Identity. Many arguments involve and depend on the notion of identity, and the logic of identity can be treated as an extension of the system of quantificational logic which we have been expounding. The formal logic of identity may be regarded as being concerned with the two-place predicate or relation, $(1)=(2)$, which has of course the meaning that (1) is identical with (2). As far as we are concerned the two argument places in this predicate may be occupied by individual symbols only: we do not consider cases of identity between X and Y where the terms X and Y denote non-individuals.

There are some intuitively obvious logically true statements or laws concerned with identity. One of these, the Law of Identity, asserts that anything is identical with itself. We may express it in the form:

$$(\forall x)[x=x].$$

Our first deductive rule concerning identity allows us to write this statement, or any other which differs from it only in the pronoun variable used, as a line anywhere in any deduction: it is not to be treated as a premiss. We call this the rule of identity. Our second and only other rule concerning identity is the rule of extensionality of identity. According to this rule if one line of a deduction is a formula, $\alpha=\alpha_1$, and another line is a formula P which contains α or α_1 in an extensional context, we may write as a new line the formula which is the same as P except that it has α_1 at any number of places where P has α or α at any number of places where P has α_1. We cannot give an entirely satisfactory account[1] of what is meant here by *extensional context* but we may say broadly that an extensional context is one which is not part of reported speech,

[1] A more comprehensive treatment of the logic of identity and descriptions will be given in a later monograph: *Identity and Descriptions*, by E. J. Lemmon.

knowledge or thought in general, and which is not concerned with the notion of meaning. For example, although Cicero is identical with Tully it would not be legitimate to replace *Cicero* by *Tully* in the context:

> Smith's exact words were, 'Cicero is an author whom I no longer read.'

This replacement would, on the other hand, properly be made in the extensional context:

> Cicero is an author whom Smith no longer reads.

Rules for identity

Rule of identity (ID)

> Any statement $(\forall \mathbf{x})[\mathbf{x}=\mathbf{x}]$ where $\mathbf{x}$ is any pronoun variable may be written as a line in any deduction.

Rule of extensionality of identity (EXT)

> $\alpha=\alpha_1$, $P \rightarrow P^{\alpha:\alpha_1}$, $P^{\alpha_1:\alpha}$, provided that any symbol for which replacement is made in P occurs in an extensional context in P.[1]

The use of these rules is illustrated in the deductions which follow. These establish the logical truth of two well-known laws of identity, the law of symmetry for identity and the law of transitivity for identity.

[1] See also reference to this rule on page 110.

ID × 1
(1) $(\forall x)[x=x]$
1 UI × 2
(2) $a_1=a_1$

Pr × 3
(3) $a_1=a_2$
2, 3 EXT × 4
(4) $a_2=a_1$

3⊢4 × 5
(5) $a_1=a_2 \supset a_2=a_1$
5 UG × 6
(6) $(\forall y)[a_1=y \supset y=a_1]$
6 UG × 7
(7) $(\forall x)(\forall y)[x=y \supset y=x]$
(Law of Symmetry)

Pr × 1
(1) $a_1=a_2$

Pr × 2
(2) $a_2=a_3$
1, 2 EXT × 3
(3) $a_1=a_3$

2⊢3 × 4
(4) $a_2=a_3 \supset a_1=a_3$

1⊢4 × 5
(5) $a_1=a_2 \supset [a_2=a_3 \supset a_1=a_3]$
5 int (VI) × 6
(6) $[a_1=a_2 . a_2=a_3] \supset a_1=a_3$
6 UG × 7
(7) $(\forall z)[[a_1=a_2 . a_2=z] \supset a_1=z]$
7 UG × 8
(8) $(\forall y)(\forall z)[[a_1=y . y=z] \supset a_1=z]$
8 UG × 9
(9) $(\forall x)(\forall y)(\forall z)[[x=y . y=z] \supset x=z]$
(Law of transitivity)

Since no premiss is undischarged at any application of *UG* in either of these deductions, a_1 and a_2 in the left-hand deduction may be any two proper names and a_1, a_2 and a_3 in the other deduction may be any three proper names.

We may, of course, if we wish, add to our two rules for identity others corresponding to the law of symmetry and the law of transitivity respectively.

8. Definite descriptions[1]; the iota operator. A common case of a statement involving identity is one in which either or both of the identified terms are definite descriptions, i.e. phrases of the form 'the so and so', e.g. 'the first man to swim the English Channel', 'the author of *Waverley*'. Since a definite description is like a name in that it normally denotes a single individual, it is natural to expect that in the application of our rules of deduction definite descriptions may be treated as if they were proper names. Consider for example the argument:

(A) The left-handed player is the former champion. Therefore anyone who can beat the left-handed player can beat the former champion.

Let us write **a** for *the left-handed player* and **b** for *the former champion* and $B(1)(2)$ for (1) *can beat* (2). The argument may now be expressed:

$\mathbf{a}=\mathbf{b}$; therefore $(\forall x)[Bx\mathbf{a} \supset Bx\mathbf{b}]$.

If **a** and **b** may legitimately be treated as though they were proper names this argument could obviously be shown to be valid by the following deduction:

Pr × 1

(1) $\mathbf{a}=\mathbf{b}$

Pr × 2

(2) $Ba_1\mathbf{a}$

2, 1 EXT × 3

(3) $Ba_1\mathbf{b}$

2⊢3 × 4

(4) $Ba_1\mathbf{a} \supset Ba_1\mathbf{b}$

4 UG × 5

(5) $(\forall x)[Bx\mathbf{a} \supset Bx\mathbf{b}]$

[1] See page 104, footnote.

We should certainly wish this particular deduction or something closely similar to it to be regarded as legitimate. There are, however, some possible complications involved in a quite general treatment of definite descriptions as though they were proper names. Some invalid arguments may be made to appear valid: for example the argument: (B) *No one passed in Latin; therefore the candidate in Latin did not pass in Latin*, is invalid, since the premiss can be true even if there was no candidate or if there were several candidates in Latin whereas the conclusion requires that there has been one candidate in Latin and one only. However, if we express the premiss as $(\forall y) \sim Ly$ and put **a** for *the candidate in Latin* and then treat **a** as if it were a proper name the use of UI enables us to deduce the conclusion $\sim L\mathbf{a}$. Again, the treating of a definite description as though it were a proper name may make a valid argument appear invalid. Consider the following argument for example:

(C) Square roots of numbers ending in 5 themselves end in 5. 765 is a number ending in 5.
Therefore, the square root of 765, if it exists, ends in 5.

If we treat the definite description *the square root of* 765 as an unanalysed proper name and use the following abbreviations: **a** for 765, $\mathbf{a}_1$ for *the square root of* 765, $V(\ )$ for ( ) *is a number ending in* 5, $R(1)(2)$ for (*1*) *is a square root of* (*2*), the argument can be expressed as follows:

(C′) $(\forall x)[Vx \supset (\forall y)[Ryx \supset Vy]$, $V\mathbf{a}$; therefore, $\mathbf{a}_1$ exists $\supset V\mathbf{a}_1$.

This argument (*C′*) is invalid: it can easily be shown that any argument form which matches it is invalid. But the original argument (C) is obviously a valid one. Hence it is clear that C′ is not a suitable way of expressing C, the reason being that this is a case in which a definite description cannot properly be treated as a proper name. We will now explain a device which can be used for handling definite descriptions in formal reasoning in place of the sometimes unreliable method of treating them as proper names.

The iota operator

Many definite descriptions can be expressed in the form: *the one and only f*, where *f* is a common noun or nominal phrase, e.g. *President of the United States.* We will explain the formal method of dealing with definite descriptions by reference to this form. Let *F*( ) stand for ( ) *is an f*. We now introduce a new notation using an inverted iota: $(\iota x)Fx$. This expression may be taken to mean *the one and only x such that Fx*, i.e. *the one and only f.* However, it has to be defined formally by means of quantificational and identity symbols already available. $(\iota x)Fx$ is an analysed singular term and we define it by defining the meanings of the simple forms of statement in which it can occur. $(\iota x)Fx$ may be used as argument to any predicate *G*( ), including the special predicate E!( ) which means ( ) *exists* and may take only an expression of this kind as argument. The two basic kinds of formula in which it may occur are $G(\iota x)Fx$ and $E!(\iota x)Fx$. These might be defined thus:

$E!(\iota x)Fx =_{df} (\exists x)[Fx.(\forall y)[Fy \supset y = x]]$, i.e. some individual is such that it is *f* and anything that is *f* is identical with it.

$G(\iota x)Fx =_{df} (\exists x)[Fx.(\forall y)[Fy \supset y = x].Gx]$, i.e. some individual is such that it is *f* and anything that is *f* is identical with it and it is *g*.

However, the *definiens* in each case may be replaced by an equivalent but more elegant formula so that we have:

3.8. Definite descriptions; the iota operator

$$E!(\iota x)Fx =_{df} (\exists x)(\forall y)[Fy \equiv y = x];$$
$$G(\iota x)Fx =_{df} (\exists x)[Gx.(\forall y)[Fy \equiv y = x]].$$

It can be seen that in accordance with these definitions if $G(\iota x)Fx$ is true then $E!(\iota x)Fx$ must be true also. Hence we must not adopt any rule of deduction which would allow us to write a formula $G(\iota x)Fx$ as a line of a deduction in a case in which $E!(\iota x)Fx$ might be false, i.e. in a case in which either no individual might be f or more than one individual might be f. For example we may not allow the rule UI to be extended unconditionally to apply to definite descriptions; for this would allow us to go from true premiss to false conclusion, a case in point being the argument (B) above if this is expressed as $(\forall y)Ny$; *therefore* $N(\iota x)Cx$, where $N(\ )$ means ( ) *did not pass in Latin* and $C(\ )$ means ( ) *was a candidate in Latin.* However, there is no objection to our allowing UI to apply to a definite description $(\iota x)Fx$ in a case in which we already have $E!(\iota x)Fx$ as a line of deduction. We may conveniently therefore add to our existing rules the rule:

(Iota 1) $(\forall \mathbf{x})P, E!(\iota \mathbf{y})F\mathbf{y} \rightarrow P^{\mathbf{x}|(\iota \mathbf{y})F\mathbf{y}}$.

If we examine the two definitions we can see also that if $E!(\iota x)Fx$ is true $F(\iota x)Fx$ must be true. This is particularly obvious if we look at the first formulations; for if $(\exists x)[Fx.(\forall y)[Fy \supset y = x]]$ is true then evidently $(\exists x)[Fx.(\forall y)[Fy \supset y = x].Fx]$ is true, but this latter formula is the *definiens* of $F(\iota x)Fx$. In view of this we may adopt the following additional rule of deduction:

(Iota 2) $E!(\iota \mathbf{x})F\mathbf{x} \rightarrow F(\iota \mathbf{x})F\mathbf{x}$.

As well as adding these new rules we may extend the applicability of the rules EG (page 58) and Ext (page 105) so as to allow them to be applied where α, α_1 are definite descriptions as well as where they are proper names or dummy names.

3.8. *Definite descriptions; the iota operator*

Our system, extended in the way described, is able to deal effectively with arguments involving definite descriptions. We illustrate this by reference to the examples mentioned earlier. The requirement of $E!(\imath \mathbf{x})F\mathbf{x}$ as a premiss for Iota 1 prevents the deduction of $N(\imath x)Cx$ from $(\forall y)Ny$ in the case of argument (B). The argument (A) on page 107 can be dealt with thus. Let us write $L(\ )$ for ( ) *is a left-handed player* and $F(\ )$ for ( ) *is a former champion.* The argument may then be expressed in this form:

$$(\imath y)Ly=(\imath y)Fy;\ \text{therefore,}\ (\forall x)[Bx(\imath y)Ly \supset Bx(\imath y)Fy].$$

If we write **a** for $(\imath y)Ly$ and **b** for $(\imath y)Fy$ we see that it can be shown to be valid by a deduction which is formally identical with that considered on page 107: this is a case in which the introduction of the iota operator has made no essential difference. In the case of (C) on the other hand, as in that of (B), it does make a difference. (C) can now be expressed as follows:

(C″) $(\forall x)[Vx \supset (\forall y)[Ryx \supset Vy]],\ V765$;
therefore, $E!(\imath z)Rz765 \supset V(\imath z)Rz765$,

and is shown to be valid in the following deduction, in which we write **a** as an abbreviation for $(\imath z)Rz765$.

Pr × 1
(1) $(\forall x)[Vx \supset (\forall y)[Ryx \supset Vy]]$

Pr × 2
(2) $V765$
1 UI × 3
(3) $V765 \supset (\forall y)[Ry765 \supset Vy]$
3, 2 MP × 4
(4) $(\forall y)[Ry765 \supset Vy]$

Pr × 5
(5) $E!\,\mathbf{a}$
4, 5 Iota 1 × 6
(6) $R\mathbf{a}765 \supset V\mathbf{a}$
5 Iota 2 × 7
(7) $R\mathbf{a}765$
6, 7 MP × 8
(8) $V\mathbf{a}$

5⊢8 × 9
(9) $E!\mathbf{a} \supset V\mathbf{a}$

In conclusion it may be noted that although the formula $a=a$, a being a proper name variable, is logically true, the formula $(\iota x)Fx=(\iota x)Fx$ is not logically true, since it is false for any $F(\ )$ such that $E!(\iota x)Fx$ is false. $(\iota x)Fx=(\iota x)Fx$ is true if but only if $E!(\iota x)Fx$ is true. Where we have $E!(\iota x)Fx$ as a line of a deduction we are able to obtain $(\iota x)Fx=(\iota x)Fx$ as follows: first we use ID to obtain, e.g. $(\forall y)[y=y]$; $(\iota x)Fx=(\iota x)Fx$ follows from this formula and $E!(\iota x)Fx$ by the rule Iota 1.

CHAPTER FOUR

1. Reformulated definition of validity; proof of invalidity. In the present chapter we shall explain how an invalid argument form may be proved to be invalid.[1] A definition of validity and invalidity in relation to quantificational argument forms has been given in chapter two, section 1. However, this definition may be expressed in an alternative form which for some of our purposes may be more convenient. We make use of the notion of the implication CI_Q corresponding to an argument form Q[2] and of the idea of the *falsifiability* of a statement form.

> A statement form F is falsifiable in a universe of discourse U if, and only if, F has at least one exemplification which is false in U.

Our reformulated definition of *validity* is:

> An argument form Q is valid if and only if there is no universe of discourse in which the corresponding implication CI_Q is falsifiable; i.e. if and only if in every universe CI_Q is non-falsifiable.

An alternative definition of logical truth in terms of falsifiability may also be given:

> A statement form is logically true if and only if there

[1] Some may find it helpful to read first the short and good treatment of this topic in I. M. Copi's *Symbolic Logic*, chapter 4, section 3.

[2] See TFL, page 52. For an argument or argument form F in which premisses are P_1, P_2, ... , P_n and conclusion is C the corresponding implication (CI_F) is $[P_1 . P_2 . \ldots . P_n] \supset C$.

is no universe of discourse in which it is falsifiable; i.e. if and only if in every universe it is non-falsifiable.

Instead of the notion of falsifiability the related notion of satisfiability is usually employed. In terms of falsifiability satisfiability may be defined thus: a statement form S is satisfiable in a universe of discourse U if and only if $\sim S$, the negation of S, is falsifiable in U.

We see from this last definition of validity that an argument form is invalid if there is at least one universe of discourse in which the corresponding implication is falsifiable. Hence an argument form may be shown to be invalid by specifying a certain universe of discourse and citing an exemplification, which is false with respect to that universe, of the implication corresponding to the form. Thus consider for example the argument form (A) $(\exists x)Fx$; *therefore* $(\forall x)Fx$. CI_A is $(\exists x)Fx \supset (\forall x)Fx$ and A may be shown to be invalid by specifying as universe of discourse the class of persons which consists of Mr. Macmillan and Mr. Gaitskell and pointing out that in this universe of discourse the following exemplification (CI_{A_1}) of CI_A is false:

(CI_{A_1}) $(\exists x)$[x was Prime Minister of the U.K. in 1961] $\supset$ $(\forall x)$[x was Prime Minister of the U.K. in 1961].

The example of a proof of invalidity which we have just given though convincing enough in its way is, from the point of view of pure logic, open to an objection, namely that its convincingness depends on factual information any or all of which may be mistaken; it is a purely contingent fact that of a given group of people one was Prime Minister in a certain period but not all were, and it is unsatisfactory that a proof of logical invalidity should depend upon contingent fact. What is desirable is that a suitable exemplification should be found in a field such as that of logic itself or of mathematics where relationships are necessary, not just contingent. Out of a multitude of possibilities we choose the following, taking as our universe of discourse the class consisting of the numbers 3, 4 and 5:

(CI_{A_2}) $(\exists x)$[x is a prime number]$\supset(\forall x)$[x is a prime number].

In the universe of discourse mentioned CI_{A_2} is false since its antecedent is true and its consequent false. Further, CI_{A_2} is not open to the objection of contingency which, as we have seen, might be brought against CI_{A_1}: from the point of view of logic prime numbers are more respectable than prime ministers.

We now have a perfectly satisfactory proof of the invalidity of the argument form A. It may be observed that we have given no explanation of how we came to think of the universe of discourse U and the statement CI_{A_2} on which the proof depends. For the sake of the proof we are not of course required to give any such explanation, since all that matters is that the universe and the statement should have the required property; and indeed it is doubtful whether in this case any very satisfactory explanation could be given. However, we have been dealing with a very simple example and it would certainly be rather unsatisfactory if in more complicated cases all that was open to us was to rummage about more or less haphazardly in the hope that eventually we might hit upon a suitable counter-example. But fortunately the situation is not quite so unfavourable. Some statement forms indeed are falsifiable only in an infinite universe of discourse, i.e. one containing an infinite number of individuals but very many falsifiable forms are falsifiable in some finite universe, and a systematic method exists by which, in theory at any rate, any such form may be shown to be falsifiable.

2. Quantifier-free equivalents. The method is made possible, as we shall see, by the fact that if we know the number of individuals in a universe of discourse which consists of a finite number of individuals we can, without knowing anything more about the universe or its elements, determine whether or not a given quantificational form Q is falsifiable in it. We shall need to refer to universes of discourse consisting of a single individual, of two individuals and so on. It will be convenient to think of these universes as all being subsets of a set of individuals which is specified to begin with. Let $\alpha_1, \alpha_2, \ldots, \alpha_i, \ldots$ be distinct individuals and let universes of discourse $U_1, U_2, \ldots, U_i, \ldots$ be constituted as follows: U_1 is the universe whose sole member is α_1; U_2 is the universe whose members are α_1 and α_2; U_3 is the universe whose members are α_1, α_2 and α_3; ... ; U_i is the universe whose members are $\alpha_1, \alpha_2, \ldots$, and α_i; .... Use will be made of the notion of an ***elementary statement form of type*** α. By an ***elementary statement form*** we mean a statement form which contains one or more predicate variables but no quantifiers. An elementary statement form is simple if it is not truth-functionally compound. We shall say that an elementary statement form is of type α if it contains no individual symbol which is not from the list of symbols: 'α_1', 'α_2', ... , 'α_i', ... , which denote respectively the distinct individuals $\alpha_1, \alpha_2, \ldots, \alpha_i$, referred to above. The following are examples of elementary statement forms of type α: (1) $F\alpha_1$; (2) $G\alpha_1\alpha_2$; (3) $F\alpha_2 \vee F\alpha_1$; (4) $[F\alpha_1 \supset [G\alpha_2\alpha_2 \vee G\alpha_1\alpha_2]] \,.\, {\sim}F\alpha_1$; (5) $F\alpha_3\alpha_2\alpha_3\alpha_1 \equiv G\alpha_3$; (6) $[{\sim}F\alpha_2 \,.\, G\alpha_2\alpha_1\alpha_2 \,.\, {\sim}G\alpha_1\alpha_2\alpha_1] \supset F\alpha_1$.

We must begin by establishing the following two propositions on the basis of which the justification of the method rests.

(I) For any quantificational statement form Q, whatever number n may be, there can be found an elementary statement form of type α which is falsifiable if and only if Q is falsifiable in the universe U_n. This elementary statement form may be known as the QFE^n of Q, i.e. the quantifier-free equivalent of Q with respect to the universe U_n.

(II) An elementary statement form of type α is falsifiable if and only if it is not a truth-table tautology.

(I) A quantificational form Q may or may not contain at least one proper name variable. We show first (i) how to find, for a form Q which contains no proper name variables, the QFE^n of Q, i.e. an elementary statement form of type α which is equivalent to Q with respect to U_n, the universe of discourse consisting of the n distinct individuals $\alpha_1, \alpha_2, \ldots, \alpha_n$.

We begin with a simple example. Consider the quantificational forms $(\forall x)Fx$ and $(\exists x)Fx$ with respect first of all to the universe U_1, consisting of the single individual α_1. If we remember the meaning of $(\forall x)Fx$ it is evident that, with respect to U_1, there is an equivalent form involving only a singular quantifier, namely $(\alpha_1 x)Fx$. This in its turn is obviously equivalent to $F\alpha_1$ which contains no quantifiers of any kind. Again with respect to U_1 $(\exists x)Fx$ is equivalent to a form which is devoid of quantifiers; for

$$(\exists x)Fx \equiv (\alpha_1 x)Fx \equiv F\alpha_1.$$

Next let us consider the same two forms in relation to the universe U_2 which consists of the two individuals α_1 and α_2. $(\forall x)Fx$ is in this case obviously equivalent to $(\alpha_1 x)Fx . (\alpha_2 x)Fx$ and thus to $F\alpha_1 . F\alpha_2$. Again it can easily be seen that $(\exists x)Fx \equiv (\alpha_1 x)Fx \vee (\alpha_2 x)Fx \equiv F\alpha_1 \vee F\alpha_2$. In general with respect to a universe U_n of n distinct individuals, $\alpha_1, \alpha_2, \ldots, \alpha_n$,

$$\begin{aligned}(\forall x)Fx &\equiv (\alpha_1 x)Fx . (\alpha_2 x)Fx . \ldots . (\alpha_n x)Fx \\ &\equiv F\alpha_1 . F\alpha_2 . \ldots . F\alpha_n\end{aligned}$$

and

$$\begin{aligned}(\exists x)Fx &\equiv (\alpha_1 x)Fx \vee (\alpha_2 x)Fx \vee \ldots \vee (\alpha_n x)Fx \\ &\equiv F\alpha_1 \vee F\alpha_2 \vee \ldots \vee F\alpha_n.\end{aligned}$$

4.2. *Quantifier-free equivalents*

Thus we see that for each of the simple quantified forms $(\forall x)Fx$ and $(\exists x)Fx$ as construed with respect to a universe U_n there is an equivalent elementary statement form of type α, i.e. a quantifier-free equivalent with respect to U_n, i.e. a QFE^n. Since this is so it is not difficult to see that for any quantificational form Q, however complex, there must be a QFE^n. Many students, however, find great difficulty in constructing a QFE^n for Q in cases where Q has any degree of complexity; and we shall now describe a straightforward procedure whereby this construction may be done.

The procedure may involve three types of operation: analysis, dequantification and synthesis. In dequantification for a formula F which is governed by a quantifier another formula is found which is free of that quantifier and yet is equivalent to F with respect to the universe in question. For the many quantificational forms which are governed by a quantifier, and contain that governing quantifier only, dequantification is the only one of the three operations which is needed. It is in fact the basic part of the procedure and we begin by showing how it may be carried out. A formula which is to be dequantified is given a reference number and the following table is used.

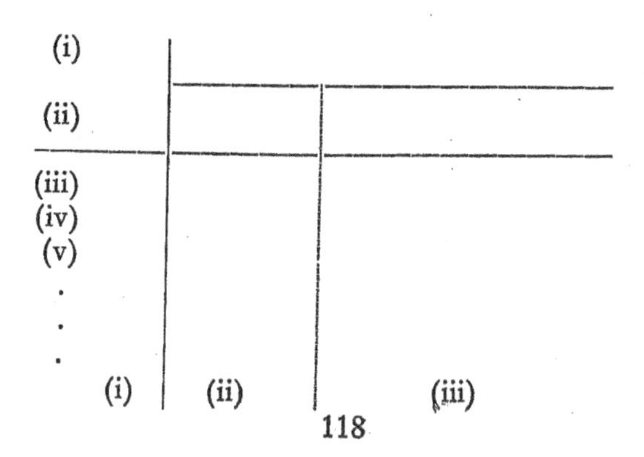

In the top line (i) we write the reference number of the formula. On the second line we write the formula itself with the governing quantifier in column (ii), to the left of the dotted line, and the matrix in column (iii); and in column (i) we write separately the pronoun variable which occurs in the initial quantifier of the formula. In column (i) starting at line (iii) we enter 'α_1', 'α_2', ... , 'α_n'. In column (ii) line (iii) is always left blank; if n is greater than 1 the other lines are filled in according to the type of the initial quantifier which appears in this column at line (ii): if the quantifier is universal we enter the conjunction sign . opposite each of 'α_2', 'α_3', ... , 'α_n'; if it is particular we enter $\vee$ in each of these places. In column (iii) at lines (iii) and following the matrix is copied down n times with the following changes: the pronoun variable which now appears in line (ii) at the top of column (i) is replaced at every occurrence in the replicae of the matrix, by 'α_1' in line (iii), by 'α_2' in line (iv), and so on, i.e. by the individual term which is already written at the left-hand side of the line in question. When this has been done for each of 'α_1', 'α_2', ... , 'α_n' we may read off the QFE^n; we do this by reading off as a single formula what is written in columns (ii) and (iii) starting with line (iii), taking the lines in order and of course ignoring the dotted line which separates columns (ii) and (iii).

As examples let us take the formulae:

(1) $(\forall x)[Fx.[Gx \supset \sim Hx]]$

and

(2) $(\exists y)[[Fy \vee Gy].Hy]$.

We fill in the tables as follows:

	1	
x	$(\forall x)$	$[Fx.\ [Gx \supset \sim Hx]]$
α_1		$[F\alpha_1.[G\alpha_1 \supset \sim H\alpha_1]]$
α_2	•	$[F\alpha_2.[G\alpha_2 \supset \sim H\alpha_2]]$
.	.	
.	.	
.	.	
α_n	•	$[F\alpha_2.[G\alpha_n \supset \sim H\alpha_n]]$

	2	
y	$(\exists y)$	$[[Fy \vee Gy].Hy]$
α_1		$[[F\alpha_1 \vee G\alpha_1].H\alpha_1]$
α_2	v	$[[F\alpha_2 \vee G\alpha_2].H\alpha_2]$
.	.	
.	.	
.	.	
α_n	v	$[[F\alpha_n \vee G\alpha_n].H\alpha_n]$

From these completed tables we can read off various QFE's. For example QFE^1 of (1) is $[F\alpha_1.[G\alpha_1 \supset \sim H\alpha_1]]$ and QFE^2 of (1) is

$$[F\alpha_1.[G\alpha_1 \supset \sim H\alpha_1]].[F\alpha_2.[G\alpha_2 \supset \sim H\alpha_2]];$$

QFE^1 of (2) is $[[F\alpha_1 \vee G\alpha_1].H\alpha_1]$. QFE^2 of (2) is

$$[[F\alpha_1 \vee G\alpha_1].H\alpha_1] \vee [[F\alpha_2 \vee G\alpha_2].H\alpha_2].$$

Of course we have to be able to deal also with formulae which are not governed by an initial quantifier or of which parts are multiply quantified. In such cases analysis and synthesis are needed as well as dequantification. We must begin by carrying out a structural analysis of the formula. A suitable method is this: immediately to the left of each quantifier draw a vertical line and at right-angles draw a line exactly covering the formula governed by the quantifier. When the analysis has been done we dequantify in turn each quantified component and at each stage fit the newly discovered QFE into its place in the structure—this is what is meant here by synthesis—until finally we have a QFE for the whole. This rather rough-and-ready description may be sufficiently clarified by some illustrations. To find QFE's for the following formula:

(1) $(\exists x) \sim Fx \,.\, (\forall x)(\forall y)[Fx \supset Gxy]$.

We first analyse (1) thus:

1		
2	3	
		4
$(\exists x) \sim Fx$.	$(\forall x)$	$(\forall y)[Fx \supset Gxy]$

The numbers are for reference only of course and it is arbitrary what number is assigned to what part. To some extent it is arbitrary also in what order dequantification is done but in dealing with a multiply quantified component we will start here with the most subordinate part.[1] In the case of (1) we will find in order the QFE's of (4), (3) and (2). It can be seen from the structural analysis that (1) is the conjunction of (2) and (3) and the QFE of (1) will also be the conjunction of the QFE's of (2)

[1] Some may regard an alternative procedure, where we begin by eliminating the outermost quantifier, as more intelligible in that at each stage we are handling statements or statement forms and never pronominal clauses. However, it can be seen that the result is the same in either case.

and (3). We shall suppose that we are interested only in the QFE^2.

	4	
y	$(\forall y)$	$[Fx \supset Gxy]$
α_1		$[Fx \supset Gx\alpha_1]$
α_2	.	$[Fx \supset Gx\alpha_2]$
QFE^2 of 4	$[Fx \supset Gx\alpha_1] . [Fx \supset Gx\alpha_2].$	

We now write the QFE^2 of (4) into (3) in the place shown in the analysis and find the QFE^2 of the reconstructed (3).

	3	
x	$(\forall x)$	$[\,[Fx \supset Gx\,\alpha_1] . [Fx \supset Gx\,\alpha_2]\,]$
α_1		$[\,[F\alpha_1 \supset G\alpha_1\alpha_1] . [F\alpha_1 \supset G\alpha_1\alpha_2]\,]$
α_2	.	$[\,[F\alpha_2 \supset G\alpha_2\alpha_1] . [F\alpha_2 \supset G\alpha_2\alpha_2]\,]$
QFE^2 of 3		$[\,[F\alpha_1 \supset G\alpha_1\alpha_1] . [F\alpha_1 \supset G\alpha_1\alpha_1]\,] .$ $[\,[F\alpha_2 \supset G\alpha_2\alpha_1] . [F\alpha_2 \supset G\alpha_2\alpha_2]\,]$

Next we find the QFE^2 of (2)

	2	
x	$(\exists x)$	$\sim Fx$
α_1		$\sim F\alpha_1$
α_2	$\vee$	$\sim F\alpha_2$
QFE^2 of 2		$\sim F\alpha_1 \vee \sim F\alpha_2$

Finally we conjoin the QFE's of (2) and (3) and thus obtain the QFE^2 of (1):

$$[\sim F\alpha_1 \vee \sim F\alpha_2].[[F\alpha_1 \supset G\alpha_1\alpha_1].[F\alpha_1 \supset G\alpha_1\alpha_2]].$$
$$[[F\alpha_2 \supset G\alpha_2\alpha_1].[F\alpha_2 \supset G\alpha_2\alpha_2]].$$

We have been dealing with the case in which Q does not contain any proper name variable. We now (ii) show how to find the QFE^n of Q in a case where Q does contain one or more proper name variables. This is a simple matter. One first obtains a formula which is known as a universal closure of Q. The QFE^n of Q is the QFE^n of any universal closure of Q. Since, as will be seen, a universal closure of a formula is necessarily free of proper name variables, the QFE^n of a universal closure of Q can be found by the method described under (i). A ***universal closure*** of a formula Q which contains m proper name variables $\mathbf{a}_1, \mathbf{a}_2, \ldots, \mathbf{a}_m$ is a formula:

$$(\forall \mathbf{x}_1)(\forall \mathbf{x}_2) \ldots (\forall \mathbf{x}_m)\, Q^{\mathbf{a}_1|\mathbf{x}_1,\, \mathbf{a}_2|\mathbf{x}_2, \ldots,\, \mathbf{a}_m|\mathbf{x}_m},$$

where $\mathbf{x}_1, \mathbf{x}_2, \ldots, \mathbf{x}_m$ are distinct pronoun variables and in each case $\mathbf{a}_i$ is free for $\mathbf{x}_i$ in Q.[1] It can be seen that any statement which is an exemplification of Q and is false, and hence can be used to prove the falsifiability of Q in a universe U_n, can also be used to prove the falsifiability of a universal closure of Q in a universe U_n; for example the form Fa_1 is shown to be falsifiable in the universe of discourse consisting of the numbers 1, 2, 3, 4, 5 by the fact that the statement *4 is a prime number* which exemplifies Fa_1 is false, and this same fact also proves the falsifiability of any universal closure of Fa_1, e.g. $(\forall x)Fx$. Further, only such a statement can serve to prove the falsifiability of Q. Hence Q is falsifiable if and only if a universal closure of Q is falsifiable. Now a universal closure of Q is of course equivalent, with respect to U_n, to its QFE^n as obtained in the way described under (i). It follows that Q is falsifiable if and only if the QFE^n of

[1] See chapter 1, section 11, last paragraph.

any universal closure of Q is falsifiable. By definition the QFE^n of Q is identical with the QFE^n of any universal closure of Q. Hence the QFE^n of Q has the required property that it is falsifiable in U_n if and only if Q is falsifiable in U_n.

As an example let us consider the formula

(1) $(\exists x)Fx \supset Fa$.

Since (1) contains a proper name variable a, to find its QFE^n, e.g. its QFE^2, we first obtain a universal closure of (1). This is (1′) $(\forall y)[(\exists x)Fx \supset Fy]$. The structural analysis of (1′) is:

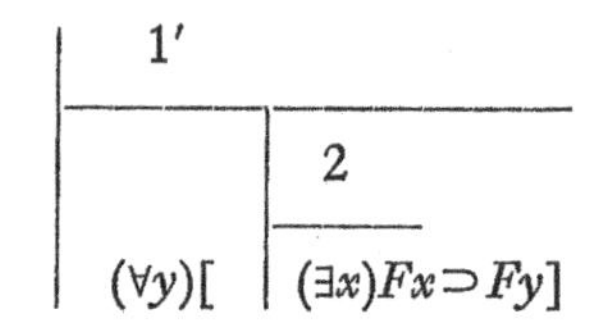

The QFE^2 of 2 is $F\alpha_1 \vee F\alpha_2$ and so the QFE^2 of the formula within square brackets is $[F\alpha_1 \vee F\alpha_2] \supset Fy$. The reformulated (1′) becomes:

$$(\forall y)[\,[F\alpha_1 \vee F\alpha_2] \supset Fy],$$

of which the QFE^2 can easily be found to be

$$[\,[F\alpha_1 \vee F\alpha_2] \supset F\alpha_1] \,.\, [\,[F\alpha_1 \vee F\alpha_2] \supset F\alpha_2].$$

This last formula being the QFE^2 of (1′) is also the required QFE^2 of (1). It is falsifiable in U_2 if and only if (1) is falsifiable in U_2.

Finally (iii) the above methods may be used for a form containing an iota expression if this is first replaced by its *definiens*.

3. Testing quantifier-free equivalents for falsifiability. (II.) An elementary statement form of type α is a truth-functional compound in which each basic component is quantifier-free and consists of a predicate variable together with arguments from the list: 'α_1', 'α_2', ..., 'α_i', ... Some examples have been given on page 116. In our method of proving invalidity it will be assumed that a statement form of this type may legitimately be tested for falsifiability by a simple truth-table method and we have now to prove that this assumption is correct.

Let us first try to make clear exactly what is in question. Consider a formula (F) which is a truth-functional compound and which has atomic truth-functional components C_1, C_2, ..., C_n. By a possible assignment A of truth-values to the components of F we shall mean a possible way of writing against each of the components C_1, C_2, ..., C_n one or other of the truth-value signs, *1*, *0*. We say that F is a truth-table tautology if and only if for every possible assignment of truth-values to C_1, C_2, ..., C_n the truth-value *1* has to be assigned to the corresponding case of F in accordance with standard truth-table procedure. Now suppose that F is a truth-functional compound which is not an elementary statement form of type α but which has as atomic components only propositional variables. We assume that F is falsifiable if and only if it is not a truth-table tautology; for example if F is the formula (6') $[\sim p . q . \sim r] \supset s$, we assume that it is falsifiable on the ground that it is not a truth-table tautology, as is shown by the fact that for the assignment, (A') $p/0$, $q/1$, $r/0$, $s/0$, (6') takes the value *0*. In this case this assumption is clearly justified. Distinct propositional variables are entirely independent of one another in the sense that the way in which one variable, say p, is replaced in the construction of an exemplification does not place any restriction on the ways in which any of the other variables may be replaced. So long therefore as we have at our disposal at least one statement which is true and at least one which is false we are able to construct exemplifications of (6'), or of any other truth-functional statement form which contains only statement variables, in accordance with any assignment of truth-values to variables that we may wish. Suppose, however, that F is an elementary statement form of type α; are we, in this case too, justified in assuming that F is falsifiable if and only if it is not a truth-table tautology? I do not think it is obvious that we are. Consider, for example,

$$(6)\ [\sim F\alpha_2 . G\alpha_2\alpha_1\alpha_2 . \sim G\alpha_1\alpha_2\alpha_1] \supset F\alpha_1,$$

an elementary statement form of type α, which is structurally similar to (6′). (6) is not a truth-table tautology as is shown by the fact that for the assignment, (*A*) $F\alpha_2/0$, $G\alpha_2\alpha_1\alpha_2/1$, $G\alpha_1\alpha_2\alpha_1/0$, $F\alpha_1/0$, (6) in accordance with the truth-table procedure must be assigned the value *0*; but that (6) is falsifiable follows from this if but only if (i) a statement exemplifying (6) can actually be constructed the components of which have truth-values corresponding to the assignment (*A*), i.e. the component replacing $F\alpha_2$ having the value *0*, the component replacing $G\alpha_2\alpha_1\alpha_2$ having the value *1* and so on, in the same way that a statement exemplifying (6′) can undoubtedly be constructed the components of which have truth-values corresponding to the assignment (*A′*). Now the components of (6) are not independent of one another in the same way that the components of (6′) were seen to be independent, since in any exemplification of (6) the component corresponding to $F\alpha_1$ must have its predicate in common with the component corresponding to $F\alpha_2$ and the component corresponding to $G\alpha_2\alpha_1\alpha_2$ must have its predicate in common with the component corresponding to $G\alpha_1\alpha_2\alpha_1$. For this reason it does not seem to be obvious that in this example the condition (i) must necessarily be satisfied. Still less is it obvious that the general assumption referred to above, namely that where F is any elementary statement form of type α it is falsifiable if and only if it is not a truth-table tautology, is justified. However, although this assumption is not obviously correct it is in fact correct and what we have now to do is to show that this is so.

4. Predicate and subject. We have seen in chapter one, section 7, that any singular statement may be regarded as the union of an n-place predicate and n arguments. However, a singular statement may be regarded, in another way, as a union of an n-place predicate and an ordered n-tuple of arguments; this ordered n-tuple of arguments may be referred to as the subject of the statement. For example, the statement:

(1) Edinburgh is north of London,

may be regarded as having for predicate (1) *is north of* (2) and for subject the ordered pair or couple of proper names: {*Edinburgh*, *London*} . {*Edinburgh*, *London*} is referred to as an ordered couple because the order of the two names is significant: {*Edinburgh*, *London*} is distinct from {*London*, *Edinburgh*}. Again the statement:

(2) Edinburgh is nearer to Aberdeen than to London,

may be regarded as having for predicate (2) *is nearer to* (1) *than to* (3) and for subject the ordered triple (or 3-tuple) {*Aberdeen*, *Edinburgh*, *London*}.

The union of an ordered n-tuple subject $\{a_1, a_2, \dots, a_n\}$ with an n-place predicate $F(1)(2) \dots (n)$ is brought about when the first name in the ordered n-tuple is written within the predicate in place of 1, the second name is written in place of 2, ... , the nth name in place of n and the brackets are removed, so that we have $Fa_1a_2 \dots a_n$. In our examples n has been greater than 1; the ordinary notion of order has of course no significance in relation to a single individual; however, it is harmless to define the expression *the ordered single or* 1-*tuple* $\{a_1\}$ as meaning simply the set with sole member a_1. With this definition our assertion that any singular statement may be regarded as the union of an n-place predicate with an ordered *n*-tuple of proper names is correct; the union of the 1-place predicate *F*( ) with $\{a_1\}$ is of course Fa_1.

Before we proceed with our main discussion one or two additional remarks about the predicate-subject analysis of singular statements may be made. (a) The predicate-subject analysis can be applied not only to unquantified but also to quantified singular statements, i.e. statements containing singular quantifiers; we need not go into this here except to remark that the way in which the union of predicate and subject is brought about to form a statement is less easily described for the case of quantified than for the case of unquantified singular statements and to warn any reader who may be interested that in the former case the order of singular terms within the ordered *n*-tuple which is subject is not necessarily the same as the order in which the corresponding singular quantifiers occur. (b) It should be obvious that a singular statement may sometimes be regarded in more than one way as a union of predicate and subject. Our example (i), for instance, might be analysed, in a formally different way from that indicated earlier, as the union of the predicate (2) *is north of* (1) with the subject {*London*, *Edinburgh*}. (c) The same name may occur more than once in an ordered *n*-tuple; thus {*Smith*, *Smith*} is a possible ordered couple.

What a simple elementary statement asserts is that an ordered *n*-tuple of individuals, denoted by the *n*-tuple of names which is the subject, has a certain property which is designated by the *n*-adic predicate. Thus *Smith is taller than Jones* asserts that the ordered 2-tuple (or pair) {*Smith*, *Jones*} has the property that the first-mentioned member is taller than the second-mentioned member, this property being designated by the 2-adic predicate (1) *is taller than* (2). Now one property which may be attributed to an *n*-tuple *N* is that of being identical with an *n*-tuple *N′*. Let us see what it means to assert that *N* is identical with *N′*. In the first place suppose that *N* is a 1-tuple $\{a_1\}$ and that N' is an 1-tuple $\{a'_1\}$; then to assert that N is identical with N' is simply to assert that the individual a_1 is identical with the individual a'_1. Secondly, suppose that N is a 2-tuple $\{a_1, a_2\}$ and that N' is a 2-tuple $\{a'_1, a'_2\}$; then to assert that N is identical with N' is to assert that a_1 is identical with a'_1 and a_2 is identical with a'_2. In general if N is an ordered n-tuple $\{a_1, a_2, \ldots, a_n\}$ and N' is an ordered n-tuple $\{a'_1, a'_2, \ldots, a'_n\}$ then the assertion that N is identical with N' is equivalent to the assertion that:

> a_1 is identical with a'_1 and a_2 is identical with a'_2 ... and a_n is identical with a'_n.

In view of this definition and of the fact that every individual is identical with itself we can see that the following two statements hold:

(i) any statement which asserts that an ordered n-tuple is identical with itself is true;

(ii) If a_1 and a_2 occur in corresponding places in two ordered n-tuples N and N', then if a_1 and a_2 are distinct (non-identical) individuals the statement that N is identical with N', since it is equivalent to a conjunction of statements one component of which is the false statement that a_1 is identical with a_2, is false.

We shall now show that:

(iii) if $\mathbf{F}_k(1)(2)\ldots(k)$ is a k-adic predicate variable and 'α_1', 'α_2', ... , 'α_m' are the names of the distinct individuals $\alpha_1, \alpha_2, \ldots, \alpha_m$ and $f_1, f_2, \ldots, f_n$ are n distinct elementary forms which have the predicate $\mathbf{F}_k(1)(2)\ldots(k)$ in common but have for subjects distinct ordered sets of m proper names, each of which is one of 'α_1', 'α_2', ... , 'α_m', then for any possible assignment of truth-values to $f_1, f_2, \ldots,$

f_n, a k-adic predicate P can be constructed such that the statements $s_1, s_2, \dots, s_n$, which result when P is substituted for $\mathbf{F}_k\,(1)\,(2)\,\dots\,(k)$ in $f_1, f_2, \dots, f_n$ respectively, possess respectively the truth-values assigned to $f_1, f_2, \dots, f_n$.

Let us use the symbol $\{(1), (2), \dots, (k)\}\,\Phi_k\langle\,\rangle$ as follows:

$\{(1), (2), \dots, (k)\}\,\Phi_k\langle\,\rangle = \{(1), (2), \dots, (k)\}$ is identical with one of the k-tuples listed within the following brackets: $\langle\quad\rangle$.

Now suppose that the conditions of (iii) hold. Let K_1 be the k-tuple which is the subject in f_1, let K_2 be the subject in f_2 and in general let K_i be the subject in f_i. Let A be any assignment of truth-values to $f_1, f_2, \dots, f_n$. We form the required predicate P_A as follows. *Case* 1. The truth-value *1* is assigned to at least one f_1. In the formula: $\{(1), (2), \dots, (k)\}\,\Phi_k\langle\,\rangle$ for each f_i to which the truth-value *1* is assigned in A we write the k-tuple K_i within the brackets to the right of Φ_k. The resulting formula is the required predicate P_A. *Case* 2. *1* is not assigned to any f_i (i.e. *O* is assigned to every f_i) in A. We write every K_i within these brackets and prefix to the whole formula the negation sign $\sim$. The resulting formula, i.e.

$$\sim\{(1), (2), \dots, (k)\}\,\Phi_k\langle K_1, K_2, \dots, K_n\rangle,$$

is P_A.

That P_A as so constructed has the required property in every case follows from the conditions of (iii) together with (i) and (ii) and our definition of

$$\{(1), (2), \dots, (k)\}\,\Phi_k\langle K_1, K_2, \dots, K_n\rangle.$$

We give some examples. *Example* 1. Let k be *1* and let $f_1, \dots, f_5$ be: $(f_1)\mathbf{F}_1\alpha_1, (f_2)\mathbf{F}_1\alpha_2, (f_3)\mathbf{F}_1\alpha_3, (f_4)\mathbf{F}_1\alpha_4, (f_5)\mathbf{F}_1\alpha_5$. Suppose that in an assignment A *1* is assigned to $\mathbf{F}_1\alpha_1$, $\mathbf{F}_1\alpha_2$ and $\mathbf{F}_1\alpha_5$ and *0* to $\mathbf{F}_1\alpha_3$ and $\mathbf{F}_1\alpha_4$. Then the required predicate P_A is $\{(\ )\}\,\Phi_k\langle\{\alpha_1\}, \{\alpha_2\}, \{\alpha_5\}\rangle$. The statements $s_1, \dots, s_5$ will be: (s_1) $\{\alpha_1\}\,\Phi_1\langle\{\alpha_1\}, \{\alpha_2\}, \{\alpha_5\}\rangle$; (s_2) $\{\alpha_2\}\,\Phi_1\langle\{\alpha_1\}, \{\alpha_2\}, \{\alpha_5\}\rangle$; (s_3) $\{\alpha_3\}\,\Phi_1\langle\{\alpha_1\}, \{\alpha_2\}, \{\alpha_5\}\rangle$; (s_4) $\{\alpha_4\}\,\Phi_1\langle\{\alpha_1\}, \{\alpha_2\}, \{\alpha_5\}\rangle$; (s_5) $\{\alpha_5\}\,\Phi_1\langle\{\alpha_1\}, \{\alpha_2\}, \{\alpha_5\}\rangle$. It is evident that, as required, s_1, s_2 and s_5 are true and s_3 and s_4 are false. *Example* 2. Let k be *3*. Let $f_1, \dots, f_4$ be: $(f_1)\mathbf{F}_3\alpha_1\alpha_2\alpha_3$, $(f_2)\mathbf{F}_3\alpha_2\alpha_1\alpha_3$, $(f_3)\mathbf{F}_3\alpha_2\alpha_3\alpha_3$, $(f_4)\mathbf{F}_3\alpha_1\alpha_1\alpha_1$. Suppose that in an assignment A *0* is assigned to f_1, f_2, and f_3, and *1* to f_4. Then the required predicate P_A is:

$$\{(1), (2), (3)\}\,\Phi_3\langle\{\alpha_1, \alpha_1, \alpha_1\}\rangle,$$

and $s_1, \dots, s_4$ are:

(s_1) $\{\alpha_1, \alpha_2, \alpha_3\}\,\Phi_3\langle\{\alpha_1, \alpha_1, \alpha_1\}\rangle$; (s_2) $\{\alpha_2, \alpha_1, \alpha_3\}\,\Phi_3\langle\{\alpha_1, \alpha_1, \alpha_1\}\rangle$; (s_3) $\{\alpha_2, \alpha_3, \alpha_3\}\,\Phi_3\langle\{\alpha_1, \alpha_1, \alpha_1\}\rangle$; (s_4) $\{\alpha_1, \alpha_1, \alpha_1\}\,\Phi_3\langle\{\alpha_1, \alpha_1, \alpha_1\}\rangle$.

If A assigned *0* to each of $f_1, \dots, f_4$, P_A would be:

$$\sim\{(1), (2), (3)\}\,\Phi_3\langle\{\alpha_1, \alpha_2, \alpha_3\}, \{\alpha_3, \alpha_1, \alpha_3\}, \{\alpha_2, \alpha_3, \alpha_3\}, \{\alpha_1, \alpha_1, \alpha_1\}\rangle$$

and, e.g., s_3 would be:

$$\sim\{\alpha_2, \alpha_3, \alpha_3\}\,\Phi_3\langle\{\alpha_1, \alpha_2, \alpha_3\}, \{\alpha_2, \alpha_1, \alpha_3\}, \{\alpha_2, \alpha_3, \alpha_3\}, \{\alpha_1, \alpha_1, \alpha_1\}\rangle.$$

Given (iii) it is clear that any elementary form f of type α may be tested for logical truth by a straightforward truth-table method. For given any assignment A of truth-values to the truth-functional components of f a statement s the components of which have the corresponding truth-values may be constructed as follows: each distinct k-adic predicate variable $\mathbf{F}_k(1)(2) \dots (k)$ in f is replaced in the way shown either by the predicate $\{(1), (2), \dots, (k)\}\,\Phi_k\langle\ \rangle$ or by the predicate $\sim\{(1), (2), \dots, (k)\}\,\Phi_k\langle\ \rangle$, the brackets to the right of Φ_k being filled in the appropriate way, to give a component of s which has the same truth-value as that assigned to the corresponding component of f under the assignment A. For example, suppose that under an assignment A the components of (6) $[\sim F\alpha_2 . G\alpha_2\alpha_1\alpha_2 . \sim G\alpha_1\alpha_2\alpha_1] \supset F\alpha_1$ are assigned truth values as follows: $F\alpha_1/0$, $F\alpha_2/0$, $G\alpha_2\alpha_1\alpha_2/1$, $G\alpha_1\alpha_2\alpha_1/0$ we construct a statement s having the required distribution of truth-values over its components as follows:

$F(\ )$ is replaced by $\sim\{(\ )\}\,\Phi_1\langle\{\alpha_1\}, \{\alpha_2\}\rangle$;
$G(1)(2)(3)$ is replaced by $\{(1), (2), (3)\}\,\Phi_3\langle\{\alpha_2, \alpha_1, \alpha_2\}\rangle$

and the required statement is:

$$[\sim \sim\{\alpha_2\}\,\Phi_1\langle\{\alpha_1\}, \{\alpha_2\}\rangle \,.\, \{\alpha_2, \alpha_1, \alpha_2\}\,\Phi_3\langle\{\alpha_2, \alpha_1, \alpha_2\}\rangle \,.\, \sim\{\alpha_1, \alpha_2, \alpha_1\}\,\Phi_3\langle\{\alpha_2, \alpha_1, \alpha_2\}\rangle] \supset \sim\{\alpha_1\}\,\Phi_1\langle\{\alpha_1\}, \{\alpha_2\}\rangle.$$

We have now shown that any elementary statement form of type α is falsifiable if it is not a truth-table tautology. Since obviously a statement form of this, or of any other kind, is not falsifiable if it is a truth-table tautology it is clear that we are justified in making the assertion (II) that an elementary statement form of type α is falsifiable if and only if it is not a truth-table tautology.

5. Special considerations relating to identity. The statement forms which we have had in mind in propositions (I) and (II) have been pure statement forms. By this is meant that they do not contain any predicates but only predicate variables; $(\forall x)[Fx \supset x$ *is brave*$]$ is an example of what we might call a mixed statement form whereas $(\forall x)[Fx \supset Gx]$ is a pure statement form. In considering questions of logical validity and invalidity we are not in general concerned with mixed but only with pure statement forms. However, an exception must be made. Since we regard arguments which make essential use of the notion of identity as part of logic we shall have to be able to test for logical truth or falsifiability statement forms which contain one or more instances of the identity predicate, $(1)=(2)$. For example, if we wish to obtain a proof that the argument form: $a=a_1$, $(\forall x)[Fx \supset Gx]$, Ga; *therefore* Fa_1, is invalid we may decide to examine the corresponding statement form:

$$[a=a_1.(\forall x)[Fx \supset Gx].Ga] \supset Fa_1$$

which is not pure since $(1)=(2)$ is not a predicate variable but a predicate.

Now proposition (I) holds not only for pure quantificational statement forms but also for forms which would be pure were it not for the presence of one or more identity components. For all such forms QFE^n's can be found in respect of any finite number n. Moreover the method of finding a QFE^n is just the same whether identity is involved or not.

Proposition (II), however, is true only of pure elementary statement forms of type α. It is not true of an elementary statement form of type α which contains an identity component. To see the reason for this let us take a very simple example. Compare the forms:

(i) $F\alpha_1 \supset [G\alpha_1\alpha_1 \vee \sim G\alpha_1\alpha_2]$,
(ii) $F\alpha_1 \supset [\alpha_1=\alpha_1 \vee \sim \alpha_1=\alpha_2]$.

The pure form (i) is not a truth-table tautology since for the case $F\alpha_1/1$, $G\alpha_1\alpha_1/0$, $G\alpha_1\alpha_2/1$ it takes the truth-value 0. Hence in accordance with our argument in support of proposition (II) it is falsifiable: an exemplification that is false can be constructed. (ii) likewise is not a truth-table tautology, in the sense that for the case $F\alpha_1/1$, $\alpha_1=\alpha_1/0$, $\alpha_1=\alpha_2/1$ it takes the truth-value 0. However, (ii) is different in that the case referred to cannot be exemplified: since we have stipulated that α_1 and α_2 are distinct individuals '$\alpha_1=\alpha_1$' is a true statement and '$\alpha_1=\alpha_2$' is a false statement, and in any exemplification of (ii) the former accordingly must have the value 1 and the latter the value *0*. $F\alpha_1$ may of course be exemplified either by a true or by a false statement. In view of such considerations it is necessary to amend proposition (II) by inserting the word 'pure' before the word 'elementary' and to assert in addition the following proposition:

> (II′) An elementary statement form of type α which contains one or more identity components is falsifiable if and only if, when the truth-value *1* is assigned to every identity component in which the two arguments are the same and the truth-value *0* is assigned to every identity component in which the two arguments are distinct, there is some possible assignment of truth-values to the other components such that, in accordance with formal truth-table procedure, the whole formula must take the value *0*.

6. Examples of invalidity proofs. We can now easily state a method whereby a quantificational statement form which is falsifiable in some finite universe can be shown to be fallacious; and this method can be used to prove the invalidity of any corresponding invalid argument form.

Given a quantificational statement form *F* we formulate in turn the QFE^1, the QFE^2, and so on, of *F*. *Case* 1. *F* does not contain any identity component. We examine each of these formulae in turn to see whether or not it is a truth-table tautology. If for some *i* it turns out that the QFE^i of *F* is not a truth-table tautology it follows, in accordance with (I) and (II) that *F* is falsifiable and so not logically true. *Case* 2. *F* contains one or more identity components. We examine in turn the QFE^1; the QFE^2, and so on. For each *i* we assign the truth-value *1* to each identity component of the QFE^i in which the arguments are the same and *0* to each identity component in which the arguments are distinct, e.g. 'α_1' and 'α_2', and we test whether there is any possible assignment of truth-values to the remaining components such that the whole QFE^i takes the value *0*. If for some i it turns out that there is a possible assignment such that the QFE^i takes the value *0* it follows in accordance with (I) and (II′) that F is falsifiable and so not logically true.

Given a quantificational argument form Q we first form the corresponding implication, CI_Q, of Q. Since CI_Q is a statement form we can use the method just described and attempt to show that it is falsifiable in some universe U_i. If this is shown it follows, in accordance with our definition of invalidity that the argument form Q is invalid.

We give now some examples.

Example 1. (7·1) (i) $(\forall x)[Fx \supset Gx]$, (ii) $(\forall x)Gx$; *therefore* (iii) $(\forall x)Fx$ is shown to be invalid as follows.

We form $CI_{7\cdot 1}$. This is:

$$[(\forall x)[Fx \supset Gx].(\forall x)Gx] \supset (\forall x)Fx.$$

We now find the QFE^1 of $CI_{7\cdot 1}$ as shown below

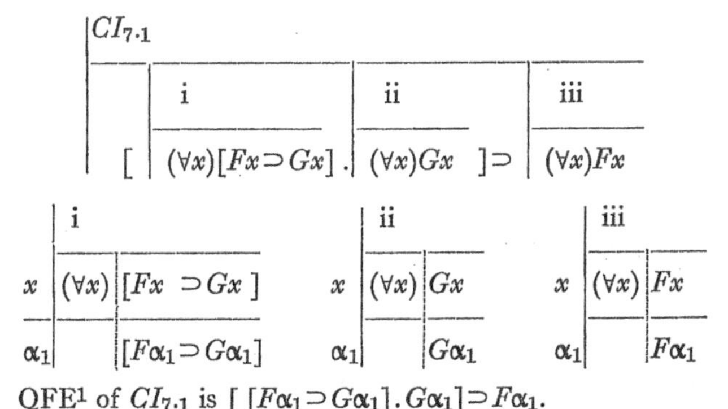

QFE^1 of $CI_{7\cdot 1}$ is $[\,[F\alpha_1 \supset G\alpha_1].G\alpha_1] \supset F\alpha_1$.

Next we use a truth-table method to discover whether or not this formula is a truth-table tautology. The case $F\alpha_1/0$, $G\alpha_1/1$ shows that it is not a truth-table tautology. (7·1) is therefore shown to be an invalid argument form.

Example 2. (19.1) $[(\exists x)Fx.(\exists x)Gx] \supset (\exists x)[Fx.Gx]$.

The QFE[1] for this formula is: $[F\alpha_1.G\alpha_1] \supset [F\alpha_1.G\alpha_1]$, which is of course a tautology. However, 19·1 is not a logical truth since although QFE[1] is tautologous it turns out that QFE[2] is not.

19.1				
	i	ii		iii
[	$(\exists x)Fx$.	$(\exists x)Gx$	] $\supset$	$(\exists x)[Fx.Gx]$

	i				ii				iii	
x	$(\exists x)$	Fx		x	$(\exists x)$	Gx		x	$(\exists x)$	$[Fx\ .\ Gx]$
α_1		$F\alpha_1$		α_1		$G\alpha_1$		α_1		$[F\alpha_1.G\alpha_1]$
α_2	v	$F\alpha_2$		α_2	v	$G\alpha_2$		α_2	v	$[F\alpha_2.G\alpha_2]$

QFE[2] of (19·1) is:

$$[\,[F\alpha_1 \vee F\alpha_2].[G\alpha_1 \vee G\alpha_2]\,] \supset [\,[F\alpha_1.G\alpha_1] \vee [F\alpha_2.G\alpha_2]\,]$$

which the case $F\alpha_1/1$, $F\alpha_2/0$, $G\alpha_1/0$, $G\alpha_2/1$, shows to be non-tautologous. (19·1) is therefore falsifiable and so fallacious.

Quantificational inference involving both truth-functors and quantifiers is liable to two types of fallacy, one resulting from a wrong use of truth-functors and the other from a wrong use of quantifiers. The first sort of fallacy can, I think, always be exposed by examining the QFE[1] for the argument form in question but the second sort can be exposed only by examining the QFE^n for the argument form in some case where $n>1$ or else, in a case in which the form is not falsifiable in any finite universe, by discovering an exemplification which is false in an infinite universe. In example 1 we have dealt with a primarily truth-functional, and in example 2 with a primarily quantificational, fallacy. Our method of proving truth-functional fallaciousness may, however, be very much simplified. The method, as we have just seen, makes use only of a QFE^1, not of any QFE^n where $n>1$. We can see by inspecting the procedure described for finding QFE's that in effect we get a QFE^1 by deleting all quantifiers and substituting 'α_1' for each pronoun variable. However, it is not difficult to see that when this is done each predicate letter, F, G, H, ... will always be followed by exactly the same sequence of letters, i.e. if F is followed at one occurrence by 'α_1' it will be followed at every occurrence by 'α_1' if it is followed at one occurrence by '$\alpha_1\alpha_1$' it will be followed at every occurrence by '$\alpha_1\alpha_1$' and so on. This being so when we are determining whether or not the QFE^1 is a tautology we may ignore the letters 'α_1', 'α_2' ..., 'α_i', ... and treat the predicate letters as if they were propositional variables. This simplified form of a QFE^1 will be referred to as an SQFE. It is found by deleting all quantifiers and all pronoun variables. Unfortunately no corresponding simplification of a QFE^n is possible where $n>1$. In example 1 the SQFE of $CI_{7\cdot1}$ may be written down at once as: $[\,[F\supset G].G]\supset F$. It is shown to be non-tautologous by the case $F/0$, $G/1$.

Example 3. (12·1) $(\forall x)[Fx\supset Gx].(\exists x)[Gx.Hx]$; *therefore* $(\exists x)[Fx.Hx]$ is shown to be invalid as follows. $CI_{12\cdot1}$ is:

$$[(\forall x)[Fx\supset Gx].(\exists x)[Gx.Hx]\,]\supset(\exists x)[Fx.Hx]$$

and the SQFE of $CI_{12\cdot1}$ is:

$$[\,[F\supset G].[G.H]\,]\supset[F.H].$$

This is proved to be non-tautologous by the case $F/0$, $G/1$, $H/1$.

Example 4. (22·1) $[(\forall x)Fx\supset(\forall x)Gx]\supset(\forall x)[Fx\supset Gx]$ has for SQFE:

$$[F \supset G] \supset [F \supset G]$$

which is of course tautologous. However, QFE² of 22·1 is found to be:

$$[[F\alpha_1 . F\alpha_2] \supset [G\alpha_1 . G\alpha_2]] \supset [[F\alpha_1 \supset G\alpha_1] . [F\alpha_2 \supset G\alpha_2]]$$

This is proved to be non-tautologous by the case: $F\alpha_1/1$, $F\alpha_2/0$, $G\alpha_1/0$, $G\alpha_2/0$. Hence (22·1) is a fallacious principle.

Example 5. (46·1) $(\forall y)(\exists x)Fxy \supset (\exists x)(\forall y)Fxy$ is a fallacious form involving multiple quantification, but invalidity can be proved without modification of our methods. The SQFE of 46·1 would of course be $F \supset F$ which is a tautology, but a proof of invalidity can be obtained by examining the QFE². We show below a structural analysis of 46·1 and the working out of the QFE² for part of the expression.

46.1				
2			4	
	1			3
$(\forall y)$	$(\exists x)Fxy$	$\supset$	$(\exists x)$	$(\forall y)Fxy$

	1			2	
x	$(\exists x)$	$Fx\ y$	y	$(\forall y)$	$[F\alpha_1 y \vee F\alpha_2 y]$
α_1		$F\alpha_1 y$	α_1		$[F\alpha_1\alpha_1 \vee F\alpha_2\alpha_1]$
α_2	$\vee$	$F\alpha_2 y$	α_2	.	$[F\alpha_1\alpha_2 \vee F\alpha_2\alpha_2]$
QFE²of1	$F\alpha_1 y \vee F\alpha_2 y$		QFE²of2	$[F\alpha_1\alpha_1 \vee F\alpha_2\alpha_1] . [F\alpha_1\alpha_2 \vee F\alpha_2\alpha_2]$	

The QFE^2 of the whole formula is found to be:

$$[[F\alpha_1\alpha_1 \vee F\alpha_2\alpha_1].[F\alpha_1\alpha_2 \vee F\alpha_2\alpha_2]] \supset [[F\alpha_1\alpha_1.F\alpha_2\alpha_1] \vee [F\alpha_1\alpha_2.F\alpha_2\alpha_2]].$$

This is proved to be non-tautologous by the case $F\alpha_1\alpha_1/1$, $F\alpha_1\alpha_2/1$, $F\alpha_2\alpha_1/0$, $F\alpha_2\alpha_2/0$.

In examples 6 and 7 we have cases of argument forms containing proper name variables and in example 8 a case of an argument form containing a propositional variable.

Example 6. (1·1) $(\forall x)[Fx \supset Gx]$, Ga; *therefore* Fa. $CI_{1\cdot1}$ is $[(\forall x)[Fx \supset Gx].Ga] \supset Fa$.
A universal closure ($CI_{1\cdot1}'$) of $CI_{1\cdot1}$ is:

$$(\forall y)[[(\forall x)[Fx \supset Gx].Gy] \supset Fy].$$

The SQFE of $CI_{1\cdot1}'$ and so of $CI_{1\cdot1}$ is:

$$[[F \supset G].G] \supset F.$$

This is shown to be non-tautologous by the case $F/0$, $G/1$. Hence 1·1 is invalid.

Example 7. (i) $[(\forall x)Fxa \supset (\forall x)Gx] \supset (\forall x)[Fxa \supset Gx]$.
A universal closure of (i) is:

$$(i')\ (\forall y)[[(\forall x)Fxy \supset (\forall x)Gx] \supset (\forall x)[Fxy \supset Gx]].$$

The SQFE of (i′) and so of (i) is:

$$[F \supset G] \supset [F \supset G],$$

which is of course a truth-table tautology.
The QFE^2 of (i′) and so of (i) is:

$$[[[F\alpha_1\alpha_1.F\alpha_2\alpha_1] \supset [G\alpha_1.G\alpha_2]] \supset [[F\alpha_1\alpha_1 \supset G\alpha_1].[F\alpha_2\alpha_1 \supset G\alpha_2]]].$$
$$[[[F\alpha_1\alpha_2.F\alpha_2\alpha_2] \supset [G\alpha_1.G\alpha_2]] \supset [[F\alpha_1\alpha_2 \supset G\alpha_1].[F\alpha_2\alpha_2 \supset G\alpha_2]]].$$

4.6. *Examples of invalidity proofs*

The case $F\alpha_1\alpha_1/1$, $F\alpha_2\alpha_1/0$, $G\alpha_1/0$ shows that the upper line conjunct and so the whole QFE² of (i) is non-tautologous. (i) is therefore not logically true.

Example 8. (35·1) $[(\forall x)Fx \supset p] \supset (\forall x)[Fx \supset p]$.
The SQFE of (35·1) is the tautology: $[F \supset p] \supset [F \supset p]$.
The QFE² of (35·1) is:

$$[\,[F\alpha_1 . F\alpha_2] \supset p] \supset [\,[F\alpha_1 \supset p] . [F\alpha_2 \supset p]].$$

In exemplifications of this formula p may be any statement and its truth-value is consequently independent of the truth-values of the statements exemplifying $F\alpha_1$ and $F\alpha_2$. There are therefore $2^3 = 8$ possible assignments of truth-values and we can see from the case $F\alpha_1/0$, $F\alpha_2/1$, $p/0$ that the QFE² of (35·1) is non-tautologous and that (35·1) itself is therefore not logically true.

In example 9 we have a statement form which contains an identity component.

Example 9. $(\exists x)(\forall y)[\,[Fx . Gx] \supset x = y]$.

The QFE¹ for this formula is $[F\alpha_1 . G\alpha_1] \supset \alpha_1 = \alpha_1$, which takes the value *1* whatever assignment of values is made to $F\alpha_1$ and $G\alpha_1$ since $\alpha_1 = \alpha_1$ must have the value *1*. It can be seen that in the QFE¹ of any form containing identity each identity component will necessarily be $\alpha_1 = \alpha_1$ and so have the truth-value *1*. As a result a useful SQFE can be devised for statement forms involving identity: we eliminate all quantifiers and individual symbols and write the truth-value symbol *1* in place of the identity sign = at each occurrence. In the present case this gives us as SQFE the formula $[F . G] \supset 1$ which of course takes the value *1* whatever assignment of values is made to F and G. However, $(\exists x)(\forall y)[\,[Fx . Gx] \supset x = y]$ is of course not logically true and this can be seen if we examine the QFE² which is:

$$[\,[F\alpha_1 . G\alpha_1] \supset \alpha_1 = \alpha_1] . [\,[F\alpha_1 . G\alpha_1] \supset \alpha_1 = \alpha_2]$$
$$\vee\, [\,[F\alpha_2 . G\alpha_2] \supset \alpha_2 = \alpha_1] . [\,[F\alpha_2 . G\alpha_2] \supset \alpha_2 = \alpha_2].$$

This formula takes the value *0* in the case $F\alpha_1/1$, $G\alpha_1/1$, $F\alpha_2/1$, $G\alpha_2/1$, the identity components necessarily of course having values as follows: $\alpha_1 = \alpha_1/1, \alpha_2 = \alpha_2/1$, $\alpha_1 = a_2/0$, $\alpha_2 = \alpha_1/0$.

7. Decision problems. In chapters two and three we described a method by which we may attempt to prove an argument form to be valid or a statement form to be logically true. In the present chapter we have considered methods of proving invalidity and falsifiability respectively. However, we have not described any purely mechanical method whereby the validity or invalidity of any quantificational argument form or the logical truth or falsifiability of any quantificational statement form may be decided. Given say a statement form *S* we may first try by deductive methods to prove it logically true. Failing in this we may attempt to show that it is falsifiable; in this attempt we might first try by the systematic procedure which we have just been describing to show that *S* is falsifiable in some finite universe U_n; growing weary of this after a time we might suspect that *S* is perhaps falsifiable only in some infinite universe: we may manage to think of some exemplification of *S* which is false with respect to some infinite universe, for example the universe consisting of the natural numbers, 1, 2, 3, ... , but we have been given no general systematic procedure for attempting to construct such an exemplification, and our more or less haphazard efforts may very well be unsuccessful. In this gloomy picture we are of course imagining the worst. The formula we are considering may be valid and we may succeed in hitting on a deduction which proves it to be so; or it may be falsifiable in some finite universe and we may persist long enough to find such a one; or, though it is falsifiable only in an infinite universe, we may be lucky enough to think of an exemplification which is false with respect to such a universe. However, despite these brighter possibilities the reader may well be dissatisfied: he may feel that what we have described as the worst possible is worse than the worst has any right to be. Is there not some mechanical procedure, however complicated, which if correctly applied must infallibly show whether or not any given quantificational statement form is logically true, in the same way that for example a truth-table procedure can enable us to decide whether any given truth-functional statement form is or is not logically true? The answer to this question is that there is no such decision procedure which is applicable to all formulae of quantificational logic without limitation; further, in accordance with a theorem of Alonzo Church (Church's theorem) it is not just that up to the present no such procedure happens to have been discovered: no such procedure can be discovered.

However, although this is the position with regard to quantificational logic in general, decision procedures are available for certain sections of the field. To mention the obvious first, truth-functional logic is part of quantificational logic and there are of course various decision procedures for purely truth-functional forms, i.e. statement forms or argument forms which contain, besides truth-functors, only statement variables. In respect of distinctively quantificational forms, however, the most important class for which there is a decision procedure available are *monadic* forms. These are forms which do not contain any n-adic predicate variable for any case of n greater than 1; for example, $[(\forall x)[Fx \supset Gx].(\exists x)Fx] \supset (\exists y)Gy$ is a monadic form, whereas $(\forall x)Fxx \supset (\exists x)Fxx$ is not monadic. Various decision procedures are known for monadic formulae; we shall describe the one which is most easily explained in relation to earlier sections of this chapter, though in general others may no doubt produce quicker results. It is known that any monadic form which is falsifiable at all is falsifiable in a universe of 2^n individuals, where n is the number of distinct predicate variables occurring in the form. Thus a suitable decision procedure for monadic quantificational forms is this: by the method described in sections 2, 3 and 4 of this chapter test the form for falsifiability in a universe of 2^n individuals (where n is as stated in the last sentence). The form is logically true if and only if it is not falsifiable in this universe.

Decision procedures are known for various other important classes of quantificational forms. For information about these the reader might consult W. and M. Kneale, *The Development of Logic* (1962), chapter twelve, section 4.

INDEX OF DEFINITIONS

References are to pages on which symbols or terms are defined or otherwise explained.

I. SYMBOLS

(*See also pages* 32–3)

II. TERMS

Index

Index

For Product Safety Concerns and Information please contact our EU representative GPSR@taylorandfrancis.com Taylor & Francis Verlag GmbH, Kaufingerstraße 24, 80331 München, Germany

Printed and bound by CPI Group (UK) Ltd, Croydon, CR0 4YY

01/05/2025

01858535-0001